SOCIO-CULTURAL PERSPECTIVES ON SCIENCE EDUCATION

Science & Technology Education Library

VOLUME 4

SCOPE

The book series *Science & Technology Education Library* provides a publication forum for scholarship in science and technology education. It aims to publish innovative books which are at the forefront of the field. Monographs as well as collections of papers will be published.

The titles published in this series are listed at the end of this volume.

Socio-Cultural Perspectives on Science Education

An International Dialogue

edited by

WILLIAM W. COBERN
Western Michigan University, U.S.A.

KLUWER ACADEMIC PUBLISHERS
DORDRECHT / BOSTON / LONDON

A C.I.P. Catalogue record for this book is available from the Library of Congress.

ISBN 0-7923-4987-3 (Hb)
ISNB 0-7923-4988-1 (Pb)

Published by Kluwer Academic Publishers,
P.O. Box 17, 3300 AA Dordrecht, The Netherlands.

Sold and distributed in North, Central and South America
by Kluwer Academic Publishers,
101 Philip Drive, Norwell, MA 02061, U.S.A.

In all other countries, sold and distributed
by Kluwer Academic Publishers,
P.O. Box 322, 3300 AH Dordrecht, The Netherlands.

Printed on acid-free paper

Printed in the Netherlands.

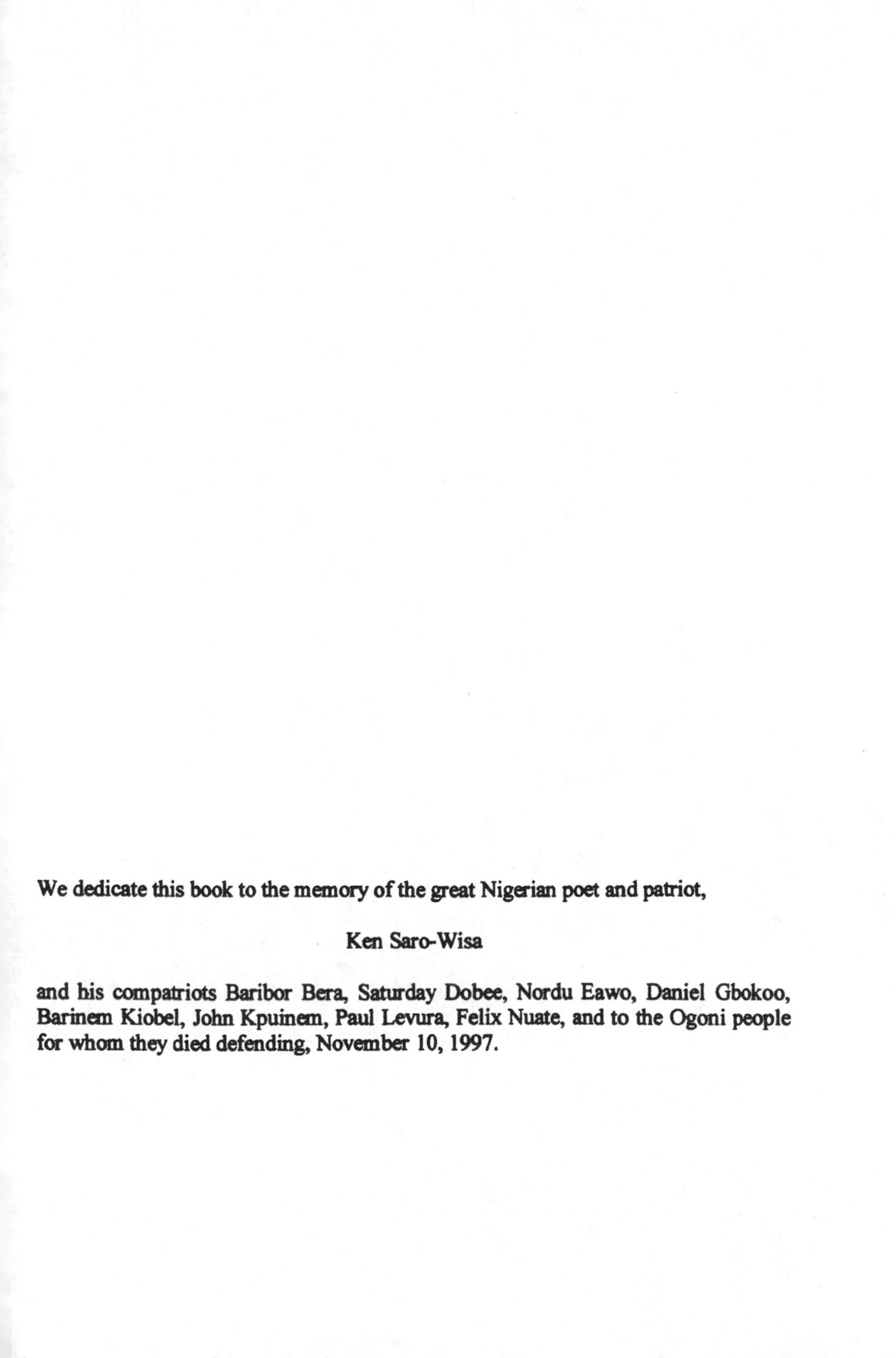

We dedicate this book to the memory of the great Nigerian poet and patriot,

Ken Saro-Wisa

and his compatriots Baribor Bera, Saturday Dobee, Nordu Eawo, Daniel Gbokoo, Barinem Kiobel, John Kpuinem, Paul Levura, Felix Nuate, and to the Ogoni people for whom they died defending, November 10, 1997.

TABLE OF CONTENTS

TABLE OF CONTENTS

Acknowledgement

We the Contributors to this volume wish to thank Zoubeida Daghar, Jacques Desautels, Patrick Dooms, Susan Edgerton, Ken Kawasaki, Elizabeth McKinley, Margaret Rutherford, and Brian Woolnough for their helpful review comments and criticisms during the process of producing this volume. As Editor, I wish personally to thank the Series Editor, Ken Tobin, for his support and encouragement from the inception of this project through to its completion.

WWC

The Cultural Study of Science and Science Education

Science is an integral part of culture. It's not this foreign thing, done by an arcane priesthood. It's one of the glories of the human intellectual tradition.
Stephen Jay Gould

We have genuflected before the god of science only to find that it has given us the atomic bomb, producing fears and anxieties that science can never mitigate.
Martin Luther King, Jr.

The reason for the book is straightforward. The profession of science education has not offered a single volume that synthesizes the research and scholarship in science studies on the purpose of science education and the effect of traditionally constituted science education on the everyday lives of people. Such a volume is needed due to the almost universal pressures to increase the level of public literacy in the sciences. The importance of science is conveyed in the above quote from paleontologist and science popularizer Stephen Jay Gould.[1] The pressures exist because of increasing economic, political, and scientific disparities between nations whose economies and technology are highly developed (DCs) and nations with lesser economic and technological development (LDs). Similarly, there are alarming disparities amongst people living within DCs that must be addressed. Thirdly, there is an almost universal problem of increasing alienation from science as suggested in the above quote from Martin Luther King, Jr. The purpose of this volume is to offer insightful commentary on these issues that will be of value to university and college professors who teach courses and conduct research on science education, science curriculum writers, governmental policy makers – and in general to all who are interested in the science/public interface.

The book raises a crucial question, Whose interests are being served by science education? Too often officials, scholars, and the public take for granted that science education serves an economic purpose. Does it really? To what extent? Who are the actual economic beneficiaries? Moreover, there is a tendency to avoid if not ignore political issues in science education. What are the power relationships in science education? What is the political influence on equity? Is science and science education all about winners? Or, are there both winners and losers? The purpose of

[1] The references for the two quotes are: Gould, Stephen Jay, *Independent* (London, 24 Jan. 1990), and King, Martin Luther, Jr. *Strength Through Love*, ch. 13 (1963).

W. W. Cobern (ed.), Socio-Cultural Perspectives on Science Education, 1–5.

science education – its goals – is typically decided by political and education officials often with little thought given to what those goals mean for the lay populace in terms of their everyday lives. In light of what people hold dearest (religion, language, culture, gender), how does science fit with the everyday lives of people? How might it be different? The essays in this book offer a *critical appraisal* of these questions and issues. The authors have avoided polemical arguments. The book, for example, is not anti-Western. On the other hand, the authors agree that among the community of nations there has been far too much uncritical copying of Western educational practices particularly in science education. The practices and results of Western science education are instructive but not simply in terms of what *to do*. They are also instructive in terms of what *not to do*. There is no virtue in repeating others' mistakes; and in the West there is no virtue in repeating one's past mistakes.

The underlying epistemology of the book is social constructivism. The authors work from the presupposition that all ideas exist in cultural context. The book is thus very much about *culture* – both the culture of science and science education, and the cultures of the people science and science education are intended to serve. The book does not argue for cultural stasis. From an evolutionary perspective one would have to argue that static cultures are also doomed cultures. The argument offered is that science education should promote neither uncritical nor unnecessary cultural change. Whether this social constructivist perspective can be extended to science itself is another question. Here the authors are split. Some write from a strong social constructivist view of science. Others, such as myself, invoke a more limited view of social constructivism. The tenor of the book is nicely captured by Gürol Irzik's comment in Chapter Eight that "we can accept that science is a socially mediated and value-laden activity without embracing the nowadays fashionable view that science is rock bottom ideology or a game of interest and power."

In Chapter One, I introduce the concept of social constructivism and argue that regardless of what one says about the nature of scientific knowledge, the learning of science must be viewed as a social construction. Thus, any science curriculum carries with it a social perspective. That social perspective, however, need not be the same everywhere science is taught, nor should it be. It may well be that the wisest thing any group of science educators can do is to rethink the strategic issues concerning science education and the answers given for the question, What are we thinking that science can do for us? Wisdom dictates that the excesses of scientism, technicism, and materialism be avoided.

In Chapter Two, Catherine Milne and Peter Taylor of Australia note that the way science and schooling are viewed affects one's perceptions of the nature and purpose of science education, especially what one considers to be important reforms for the next century. For much of the twentieth century, the Western image of science has been dominated by an objectivist epistemology and with schooling regarded as an uncritical reproduction of the status quo of Western society. From a perspective that combines constructivist epistemology and critical theory, Milne and Taylor examine the reasons these images of science and schooling are outdated and harmful.

Israeli sociologist Gili S. Drori, in Chapter Three, notes that in Western society science is placed on a pedestal and defined in terms of national economic development. In spite of the wide spread acceptance of this *science-for-development* conceptual model as a basis for national policy, social science research provides weak support for these policy assumptions. Drori presents four arguments that challenge the model and specifies potential consequences of the national commitment to the *science-for-development* model. She concludes with a call for the reassessment of national policies concerning science and science education.

In Part I of Chapter Four, Prem Naidoo and Mike Savage (from South Africa and Malawi, respectively) draw upon experiences in both pre and post apartheid South Africa to illustrate the significant impact government science and science education policy can have on the amelioration of racism and the promotion of economic equity. They vigorously argue that a concept such as equity is steeped in ideology and generates conflict. It is a political concept that can be applied to any human enterprise. Even within education equity has many political interpretations. The concept of equity is part of a larger social context that is constantly shifting and is itself subject to ideological conflict. Thus concepts of equity continually change and context defines their meaning.

In Part II of Chapter Four, Kopano Taoli who is with the Foundation for Research and Development in South Africa offers a more historical account of apartheid policies and the current institutional efforts to redress the inequities left behind by the years of apartheid. He goes on to offer a sober account of the outlook for success and the obstacles that must be overcome.

The theme of equity is further developed in Chapter Five. Australian science educator Kathryn Scantlebury examines gender issues from a social constructivist perspective. She argues that one's sense of gender originates early in life and usually remains along society's strong gender-role stereotypes. This engendered knowledge can impact a teacher's behavior in the way a teacher interacts with students, as well as the students' perceptions of teachers and their own ability to learn. Similarly for science educators, constructed views of gender influence the research questions one asks, teaching practices and curricular choices. Scantlebury asks, What questions would be posed if gender were to be taken seriously? If science educators can begin to re-define our own field to recognize the important role that gender plays in the teaching and learning of science, and the variety of meanings for any concept, we may possibly begin to dissolve the strong connections between masculinity, femininity and science — and achieve "science for all".

Historically science education has been intimately associated with the European languages and yet today the vast majority of students studying science are not native speakers of European languages. In Chapter Six, Marissa Rollnick of South Africa explores the issue of language learning in science as it relates to second language learners. She seeks to illuminate the complexity of the language question and the extent to which it is intricately woven into other issues like culture, cognition and educational background. The ideas of Vygotsky provide a useful theoretical basis for understanding these connections. His idea of language as a mediator between thought and action harmonize well with constructivist conceptions of creating structures which closely match what the learner already knows. Rollnick draws upon her experience in South Africa where English is the foreign language for many students.

Chapter Seven is written from a non-Western perspective. Masakata Ogawa examines the Japanese science education program called "Rika" as a case study of a culturally informed science education curriculum. The story of "Rika" begins with the conflict between a traditional culture and an imported western culture: "Shizen", the Japanese love of Nature, versus "Nature", the Western concept of the natural, material world. Ogawa examines the Japanese philosophy of "Shizen" and how it has been incorporated into a science education program. By offering an "Epic description" of Japanese culture Ogawa is suggesting that people should capitalize upon their traditional cultures rather than shun them as outmoded in this modern age. Everyone should ask, *What should science education be like for us?*

In addition to language and culture, religion is a powerful influence on learning. In Chapter Eight, Turkish philosopher Gürol Irzik argues that because the epistemological and political dimensions of science are inseparable, teaching philosophy of science is not merely a self-subsistent epistemological activity, but at the same time a political one. This thesis is illustrated in the context of Turkey, perhaps the only secular country almost 95% of whose population consists of Moslems, with attention drawn to some problems in teaching philosophy of science in such a country. Irzik points out a viable alternative to positivist, social constructivist and postmodernist conceptions of science. Between these extremes is a healthy approach that recognizes that science is one institution among others in a society and that scientific activity is always carried out in a social context.

Writing from a Judeo-Christian perspective in Chapter Nine, Michael Poole of the United Kingdom notes that that as modern science moves further and further away from a common sense understanding of reality, the more difficult it becomes for the majority of students to view science as anything more than a calculating device. Indeed, it has been argued that the single greatest challenge facing science education today is the loss of meaning that students can construct for science. Poole examines the impact of Judeo-Christian philosophical and religious concepts on the meaning of science vis-à-vis the impact of science on meaning and values.

In the Chapter Ten, Taylor and Cobern conclude the book with a summing up of the issues that can actually be read first as a kind of advance organizer for the entire volume. Chapter Ten lays out in summary, yet critical fashion, the issues that are addressed in detail in each of the various chapters. What emerges from this book and what can profitably be kept in mind as one reads, is that science education needs to proceed within a bimodal framework. (1) Science teaching and learning occurs within a social and cultural framework or context and thus it is critical that teachers and students understand science within a context of culturally grounded values and practices. (2) For all its great power, there is within science a social component and sometimes the ideology of those who control curriculum promote a view of science more objective and powerful than is warranted. This bimodal perspective retains the significant power of science as a way of knowing while avoiding the alienating affects of traditional Western monolithic views of science and science education.

The subtitle of this book is "An International Dialogue". We the contributors view the book as *international* in that the authors come from a variety of countries and cultures. Though we are skeptical that there are many, if any, truly international solutions to the various problems in science education, we are quite certain that the problems themselves have an inter-national flavor. There are, for example, language problems in American science education as there are in South Africa. The solutions may vary but surely American science educators can benefit from a dialogue with South African expertise on the issues – so it is as well with other countries, cultures and issues.

And what do we mean by "dialogue"? We are hopeful that readers will enter into a dialogue with this volume and with each other. We have in mind what physicist David Bohm[2] called a *dialogue*. "The image this derivation [of dialogue] suggests is of a *stream of meaning* flowing among us... a flow of meaning in the whole group, out of which will emerge some new understanding... When everybody is sensitive to all the nuances going around, and not merely to what is happening in one's own mind, there forms a meaning which is shared."

---

[2] Bohm, D. (1992). On dialogue. *Noetic Sciences Review*, (23) 16-18.

William W. Cobern

Chapter 1

Science and a Social Constructivist View of Science Education[3]

One has to recognise that as the culture and substance of science spreads more widely into the popular - majority - culture, it must also *itself* undergo enrichment and de-formalization, getting cross-connected with the familiar phenomena of everyday life; and the familiar 'common sense' ideas not suppressed or declared wrong, but reconnected and re-constructed.
David Hawkins (1992)

Educators have long viewed science as either a culture in its own right or as transcending culture. More recently many educators have come to see science as one of several aspects *of* culture. In this view it is appropriate to speak of *Western* science since the West is the historic home of modern science, modern in the sense of a hypothetical-deductive, experimental approach to science. If "science" is taken to mean the casual study of nature by simple observation, then of course all cultures in all times have had their own science. There is, however, adequate reason to distinguish this view of science from modern science. It follows that science *education* is an aspect of culture and thus it is appropriate to speak of Western science education. Since the late 1970s the education literature has shown "a growing awareness that, for science education to be effective, it must take much more explicit account of the cultural context of the society which provides its setting, and whose needs it exists to serve" (Wilson, 1981, p. 29). This suggests that a simple transfer of Western educational practices to other cultures including sub-cultures within the West will not do. Indeed, statistics indicate that "far more children study science in developing countries than earlier but the evidence suggests that the great majority do not master more than a small proportion of the goals set for them" (Lewin, 1990, p. 1; also see Lewin, 1993). Moreover, the increasing pluralism within Western societies – not to mention increasing disinterest in science among students in Western societies – suggest that even within the West it is important for science educators to understand the fundamental, culturally based beliefs about the world that students and teachers bring to class, and how these beliefs are supported by culture; because, science education is successful only to the extent that science can

[3] This chapter is based on Cobern (1996a) and used here by permission from the editor of IJSE.

W. W. Cobern (ed.), Socio-Cultural Perspectives on Science Education, 7–23.

find a niche in the cognitive and cultural milieu of students. As David Hawkins points out, science must be re-connected with everyday life.

Educators tend to focus solely on the careful explication of scientific concepts, the *domestic affairs* of science education leading to the view that science curricula are readily transferable across Western borders or across sub-cultural borders within the West. Educators must also grapple with how to help students make sense of science concepts that are often quite foreign. This *foreign affairs* focus, as so aptly put by Hills (1989), is based on two premises. First, all science exists in cultural context, and second, the teaching and learning of science is often a cross-cultural activity. Here it is helpful to use Geertz' (1973, p. 5) definition of culture, "man is an animal suspended in webs of significance he himself has spun, I take culture to be those webs, and the analysis of it is not an experimental science in search of law but an interpretative one in search of meaning." Thus, science makes more than scientific sense to a scientist. It makes sense within the scientist's entire view of reality and significance. A classroom lesson seeks to make scientific sense of a scientific concept for students, but this becomes a cross-cultural activity when the scientific sense does not automatically fit with the student's more global view of reality. One would think then that the further students are culturally removed from the West the more seriously one ought to address the relevance of culture in science education. Indeed this is a concern for many science educators (see Aikenhead, 1996; Cobern & Aikenhead, 1997); and, again, it also is a concern within the West because of growing cultural pluralism.

Curriculum planners, however, have historically followed the rule of thumb that science education ought to be as much like science proper as is possible. So if science is viewed as a singularity, then so should be the view of science education. Figure 1 is a schematic representation of a strict empirical view of science. Note that science is shown in the natural world with no connection to the social world. In other words, in this view, which might be called the received view, society has little or no impact on science. Science in the main is a reflection of the natural world. Please note that at the university one can find courses on the comparative study of religion, of art, of politics, of economics – but there are no courses on the comparative study of science. It is often said that there is one nature and therefore only one science.

One Nature, One Science?

Is it ever appropriate to modify the noun *science* by placing an adjective before it? Clearly, there is little disagreement about referring to the *natural* sciences, the *physical* sciences, and the like, but what about *African* science or *feminist* science? Is there a plurality of sciences in this sense? It is a commonly held opinion that "Only one science exits, and it has no built-in point of view. The weight of the evidence determines the conclusions, whatever they may be" (Shapiro, 1986, p. 256). From this perspective, science is above culture. Cultures vary from location to location, but science transcends culture and is constant across all. It may be that idiosyncrasies and cultural variations influence meaning in art and religion, but not science. In his discussion of the unity of science, a discussion mentioning scientists *only* from Europe, C. P. Snow (1961, p. 258) said of the great physicist Rutherford, "For him the world of science was a world that lived on a plane above the nation-state..."

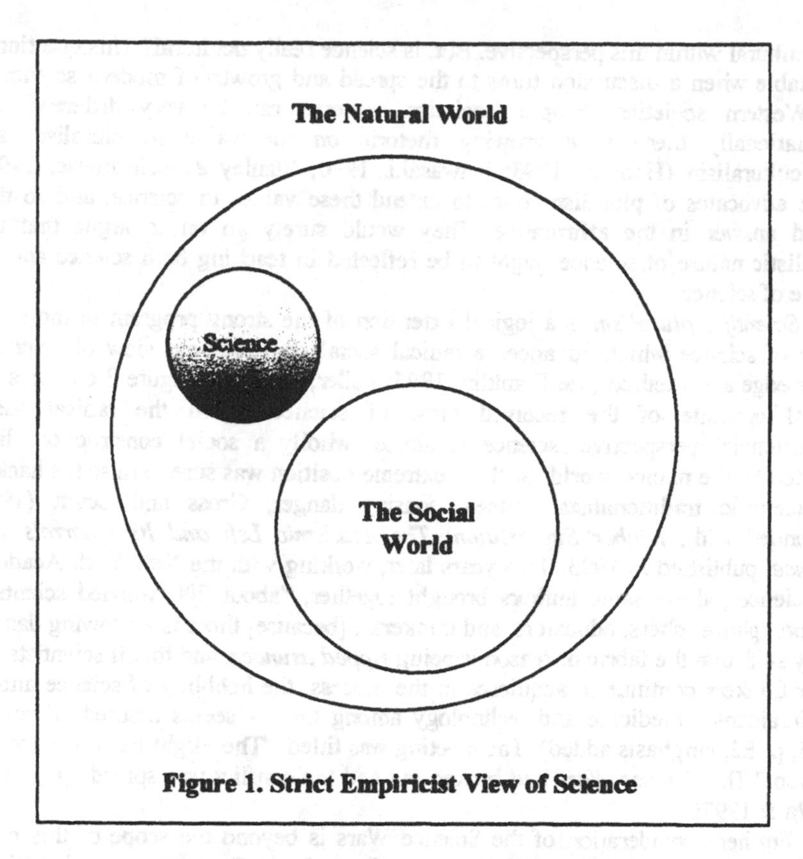

Figure 1. Strict Empiricist View of Science

This "above the nation-state" or trans-cultural position is implicit in the literature on the transfer of scientific knowledge from modern Western nations to developing, non-Western nations.

> *Modern*, ecumenical, and universal science, in other words, the science of the Scientific Revolution, originated only in Europe during the Renaissance.... During the past three centuries this has spread throughout the world, and there is no one, of whatever race, sex, colour, or creed, who cannot use it and add to it if once trained in it. (Needham, et al., 1986, p. xviii)

The eminent Pakistani physicist, Abdus Salam, was adamant that for developing nations "there is only one path to gaining ascendancy in science and technology - master science as a whole. These societies are not seduced by the slogans of 'Japanese' or 'Chinese' or 'Indian' science" (Salam, 1984, p. 285).[4] Within this perspective, a primary concern of science educators is the impediment that deficient, indigenous or alleged non scientific cultures pose to the introduction of science - which coming *from the West* is not *of the West*. Both science and science education

[4] In Chapter 8, Gürol Irzik provides insight on both moderate and radical Islamic viewpoints on science and Islam, both of which differ from Salam's rather conservative view – conservative as in scientifically conservative. Similarly, in Chapter 7, Ogawa offers a Japanese perspective on science education including what he calls "Japanized" science.

are *a*cultural within this perspective. But, is science really *a*cultural? This question is inevitable when a discussion turns to the spread and growth of modern science in non-Western societies. People, societies, cultures can be very different, and internationally there is a growing rhetoric on the value of pluralism and multiculturalism (Hodson, 1993; Kawasaki, 1996; Stanley & Brickhouse, 1994). Some advocates of pluralism want to extend these values to science, and so they would answer in the affirmative. They would surely go on to argue that this pluralistic nature of science ought to be reflected in teaching both science and the nature of science.

Scientific pluralism is a logical extension of the strong program in the social study of science which advances a radical social constructivist view of scientific knowledge and method (see Franklin, 1995; Fuller, 1991). In Figure 2 one sees the virtual opposite of the received view of science. From the radical social constructivist perspective, science is almost wholly a social construction, little affected by the natural world. Such an extreme position was sure to raise the hackles of scientific traditionalists. Indeed, sensing danger, Gross and Levitt (1993) responded with, *Higher Superstition: The Academic Left and Its Quarrels with Science*, published in 1993. Two years later, working with the New York Academy of Sciences, these same authors brought together, "about 200 worried scientists, doctors, philosophers, educators, and thinkers... [because] there is a growing danger, many said, that the fabric of reason is being *ripped asunder*, and that if scientists and other thinkers continue to acquiesce in the process, the hobbling of science and its handmaidens - medicine and technology among them - seems assured" (Browne, 1995, p. E2, emphasis added). The meeting was titled, "The Flight from Science and Reason." The *Science Wars* had broken out and the conflict was spreading (Nature, 1997a & 1997b).

Further consideration of the Science Wars is beyond the scope of this book. Suffice it to say that the social constructivists have done everyone in science education a good turn by refocusing attention on critical cultural issues with regard to knowledge. In the mid 1990s it is much easier to conceive of science learning and teaching as a cultural enterprise (see Figure 3). And, if one adopts a strong social constructivist view of science education then it becomes sensible to think about alternative constructions of science education. Before I go on to do so, however, I want to proceed with the topic of science education based on science as a singularity. In my view this has resulted in two serious errors for science education. One error is the notion of cultural deficit and the other is the notion of a non-rational mind or rationality gap.

Cultural Deficit Theory

The rise of modern science occurred first in Europe and expanded with European culture to what today is the Western world. As a way of denoting geographical location, one may accurately speak of *Western* science. This says nothing about culture, only geography. Indeed, the empiricist ideal mitigates against any consideration of a cultural influence within science. The empiricist ideal is one of strict objectivism. From this perspective, the goal of science is the complete description of the physical world independent of culture. Moreover, science is independent of persons except that an individual must have a requisite educational background of scientific and mathematical knowledge. Westerners have not thought this a *Western* view per se, but simply an accurate and appropriate view of modern science (modern as opposed to Medieval and ancient science). Many beyond Western borders agree (e.g., Salam, 1984).

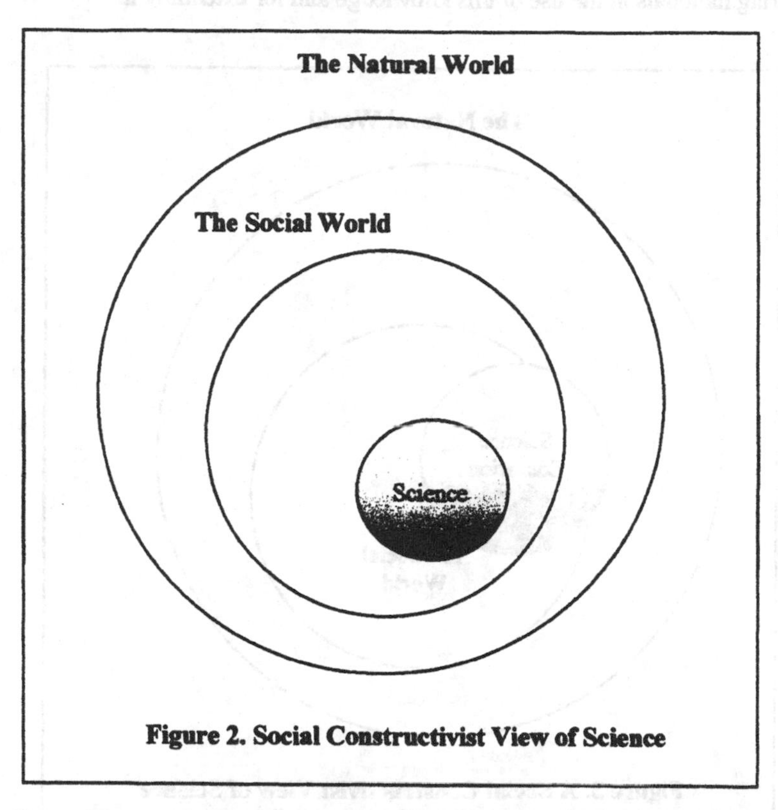

Figure 2. Social Constructivist View of Science

Except for Japan, the spread of modern science beyond Western borders began in earnest with the end of the colonial period. Independence inaugurated a period of feverish developmental activity as newly independent states of the Third World sought to improve quality of life and close the economic gap between themselves and the West. The wisdom of the day argued, "From... the present state of interaction between scientific research and society, and from the definitions of underdeveloped countries and of cultural revolution..., there follows one basic conclusion on national policy: The building of scientific research in the less developed countries into a social force relatively as strong as it is in the developed countries must have, from the first, a priority as high as, for example, economic development" (Dedijer, 1962, p. 783). Rostow's (1971) seminal book on development, *The Stages of Economic Growth*, carried this message of science for economic development to an eager audience. In the main, the way in which governments attempted to implement this perspective on science and economics was to set up a pipeline, so to speak, for transferring scientific and technological knowledge from the West to the developing nations. There was no need to begin an indigenous scientific enterprise from the ground up when one could import a highly advanced corpus of scientific knowledge from abroad. And, of course, science education – also to be imported – was the key to preparing nationals in the use of this knowledge and for extending it.

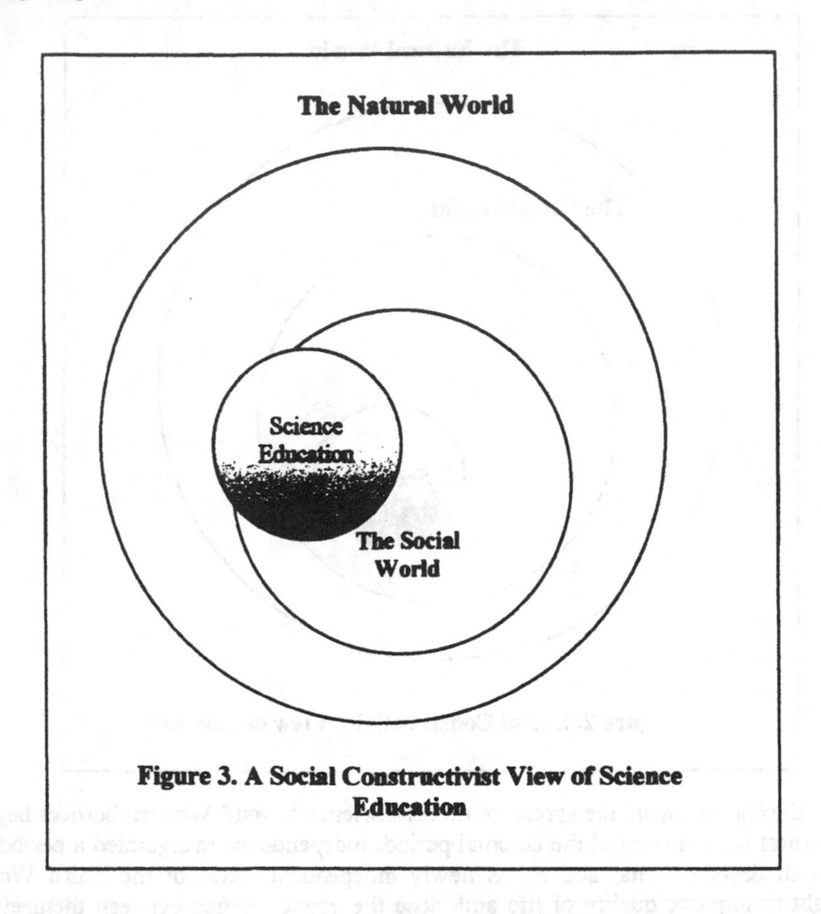

Figure 3. A Social Constructivist View of Science Education

Most development experts from the 1950s through the early 1970s embraced theories of *cultural deficit.* A deficit perspective was implicit in Basalla's (1967) seminal work on the spread of modern science to pre scientific societies. He argued that scientists coming from societies with established science (e.g., the USA or UK) in effect colonized science in new locations. Then, given certain necessary cultural, social, and economic developments, modern science would take root and take off in the now formerly pre scientific society. He argued that "resistance to science on the basis of philosophical and religious beliefs must be overcome and *replaced* by positive encouragement of scientific research. Such resistance... must be *eradicated* when science seeks a broad base of support at home" (Basalla, 1967, p. 617).

Modern science would then bring these pre scientific societies into the modern world. Moreover, many Western development experts in the 1960s believed that science would help bring about a world culture. Dedijer (1962, p. 783) wrote,

> definitions of culture are based on the hypothesis that at present *all cultures are evolving in a planned way toward a common world culture*... a global civilisation. Development then consists in carrying out with the aid of the outside world, but primarily by their own forces, a planned, rapid and simultaneous change of most complexes of their existing cultures in the general direction of the developing world culture.[5]

In the mid 1990s, with ethnic and nationalistic rivalries breaking out all over, it is hard to take such ideas seriously, but in 1962 the idea of cultural development was simply consistent with practice in the West. Western educators couched issues concerning Western minority groups, such as African-Americans, in a framework of cultural deficit theory. Moreover, development experts found support for this view in cultural anthropology. Based on Levy-Bruhl's (1926) work, *How Natives Think*, many anthropologists believed that a great divide separated modern thinking (i.e., the way modern Western intellectuals think) from primitive thinking (i.e., all forms of traditional thinking including those within the West). This viewpoint made education all the more important, because it was education that would transform primitive thinkers into modern thinkers. The underlying assumption is that some cultures promote rational thought while others do not.

In time, the deficit model lost much of its luster, due in no small way to its bald faced ethnocentricity. Nevertheless, the underlying assumption of deficit theory has survived and frequently reappears in the literature of education to day. For example, in the USA researchers periodically imply that conservative religious people have not reached a level of abstract thinking needed to understand science (e.g., Lawson & Weser, 1990). Similarly, it is not uncommon to find research that implies that non-Western, traditional people have not reached a level of abstract thinking needed to understand science (Obekubola & Jegede, 1990). In both cases, whether or not the researchers are aware of it, the research assumes a deficit model of culture. The traditional culture is deficient because it fosters and sustains modes of thought considered irrational by the researchers. In this perspective, a primary objective of science education is to inculcate scientific thinking as an antidote for irrationalism, that is, traditional thinking.

Rationality is *Not* the Issue

This view has long been held. The science historian George Basalla wrote that the establishment of independent, indigenous science in a society new to modern science requires that "resistance to science on the basis of philosophical and religious beliefs... be overcome and replaced by positive encouragement of scientific research" (1967, p. 617). "It is difficult," according to H. E. Poole (1968, p. 57), "to see how the less advanced societies can achieve the high living standards at which they aim without assimilating large portions of the Western conceptual system, not least those concepts of scientific significance." Musgrove (1982) not only concurred,

---

[5] See Dissanayoke (1990) and Vedin (1990) for an introduction to the issues of cultural homogenization driven by the pressures of technical globalization.

he further argued that a successful education lifts children out of their culture. Such opinions reflect an unnecessary "deficit" or "culture-clash" view of culture where traditional cultures are not simply viewed as different, but tacitly assumed to be less rational than modern Western culture. To date, Piagetian developmental theory has been the framework of choice for most cross-cultural research on thinking, rationality, and cognitive development. Piaget designed a set of clinical interviews based on formal propositions of logic. He inferred levels of cognitive development from performance on interview tasks. For use in education research, others have designed paper-and-pen assessments of reasoning ability based on Piaget's clinical interview procedures. In either case the inference is that a person is logical if he or she can successfully complete the assessment tasks. It is well known, however, non-Westerners typically do not perform as well as Westerners on Piagetian based tests of reasoning ability. The problem stems from a paradoxical relationship between logic and understanding.

These devices require a rather problematic assumption. To assess reasoning ability the researcher must first assume that the subject being assessed correctly understands the premises of the assessment procedures. That a person understands a task or statement, however, can only be determined by observing that the person agrees or disagrees as to (1) what statements are the same as the given one, (2) what the given statement implies, (3) what other statements are contradicted by the given statement, and (4) what is irrelevant to the given statement (see Smedslund, 1970). This is exactly where the difficulty lies. If a person cannot think well (if a person is not very rational) that person will not be able to recognize statements that are equivalent in meaning, statements that contradict each other, the implications of a statement, or the things that are irrelevant to the presented statement or problem. Then, if the subject fails to answer the test or interview question properly, the researcher is actually unable to determine whether the subject failed due to a lack of logical skill or due to a failure to understand the premises of the task. The point is that in ordinary situations it is exceedingly difficult, if not impossible, to separate rationality from comprehension. The cognitive development researcher can only proceed by assuming that a subject understands what is going on in the test. But, this is a counter-intuitive and counter experiential assumption. "In conversation we always assume that the other person is logical... When our expectations are not fulfilled, we normally attribute it to a lack of understanding... but not to genuine illogic on his part... logic must be presupposed, since it is characteristic of any activity of any integrated system and is a part of the very notion of a person" (Smedslund, 1970, p. 217-218).

Smedslund's point is that the researcher is making an assumption that defies commonsense. In effect, the research assumes what commonsense tells us *not* to assume. Abiola of Nigeria noted that, “In many investigations, including those conducted by Africans, the imported research instruments... have been taken out of the conceptual context in which they were developed... If you use a culture-bound normative instrument, you end up with a better/worse comparison inference or ‘explanation’; in most cases it is worse” (1971, p. 63). Indeed, Westerners have long evaluated other cultures by measuring their science and technology against the standards of Western science and technology with predictable results (see Adas,

1989). For education research based on Western instruments to show more positive results, the Non-western subjects would need to have acquired the particular understanding assumed by the Western oriented theory. That this will happen, is the implicit assumption under girding the straight transfer or minimal adaptation of science curricula. Students must conform to a particular understanding to avoid the label of irrationality. For example, in a recent African study the researchers concluded, "rural communities... are apt to explain naturally occurring phenomena through *irrational* means" (Okebukola & Jegede, 1990, p. 666, emphasis added). The rural students in the study were deemed irrational because of low scores on a Western based measure of reasoning ability. To no surprise the researchers found that the students held illogical ideas. The point it this: The "above culture" view of science and science education leads to notions of cultural deficit based on an assumed rationality gap. The purpose of science education from this perspective is not only the teaching of scientific thinking but also the teaching of rationality by clearing out all traditional irrational thinking. To accomplish this task policy makers employed the transfer and minimal adaptation of foreign science curricula to developing countries.

Cultural Issues and Curricular Adaptation

Science educators have called the decade of the 1960s the "golden era" of North American and European science education curriculum development (Garrison & Bentley, 1990, p. 188). The curriculum developments are called a "revolution in science education" (Prather, 1990, p. 12). UNESCO and other governmental agencies arranged for the transfer of many of these curriculum developments to Non-western, Third World nations to aid technological development and modernization. The prevailing attitude through the 1960s toward the transfer of scientific knowledge was little concerned with culture. As one Western educator put it, "a literate nation is provided with the means for substituting scientific explanations of everyday events - such as death, disease, and disaster – for the supernatural, non-scientific explanations which prevail in developing societies..." Lord (1958, p. 340). Through education, modern science is brought into developing societies where it can displace non-scientific ideas. However, from the time transfer efforts began, expatriate teachers in Third World schools, expatriate teacher education professors preparing nationals to be teachers of science, and many, many nationals with broad cross-cultural experience discovered that transferring a science education curriculum is one thing. Employing it with desired effect is quite another.

Early on, educators noted that transfer should not occur without first adapting curricula for the receiving country. Curricula, "would have to be adapted to remove an American cultural bias, for example, and substitute an African cultural bias" (Champagne & Saltman, 1964, p. 1). However, all too often educators held naive views about adaptation. Cultural adaptation simply meant changing to "terms of tropical ecology and meteorology, and increased rates of reaction in the warmer climate [and substituting] Lagos for London, *cedis* for dollars, mangoes for apples" (Wilson, 1981, p. 27). On this point, it is worth quoting at length from Dart & Pradhan (1967, p. 649) who conducted science education research in Asia:

> **Science education, in any country, is certainly a systematic and sustained attempt at communication about nature between a scientific and a nonscientific, or a partially scientific, community, and as**

such it should be particularly sensitive to the attitudes and presuppositions of both the scientist and the student. In fact, however, the teaching of science is often singularly insensitive to the intellectual environment of the students, particularly so in developing countries, where the science courses were usually developed in a foreign country and have undergone little if any modification in the process of export. Why should we suppose that a program of instruction in botany, say, which is well designed for British children, familiar with an English countryside and English ways of thinking and writing, will prove equally effective for boys and girls in a Malayan village? It is not merely that the plants and their ecology are different in Malaya; more important is the fact that the *children* and *their* ecology are also different.

Dart & Pradhan (1967) recognized a need for cultural sensitivity that goes far beyond the simple substitution of examples. In 1981, based on his extensive research in Papua New Guinea, Maddock called for a more anthropological approach to science education. Yet, six years later, Urevbu (1987) in Africa observed that concerns about cultural sensitivity in science education had yet to be heeded. One finds this observation repeated even today. At the end of the 1980s indigenisation had succeeded to the point that most former colonial nations had developed their own science programs, but too often indigenisation has meant the superficial adaptation of essentially Western curricula. The profession *still* has "a long way to go in developing ways of representing science that are not foreign, expert, and culturally unsympathetic" (Lewin, 1990, p. 18). Lest Westerners think this is a foreign problem, one should consider the possible reasons for the rising tide of anti-science sentiment within the West referred to earlier in this chapter.[6]

Science textbooks from around the globe remain strikingly similar (Altbach, 1987; Apple, 1992; McEneaney, 1995). Some similarity is to be expected. For example, one expects a discussion of the observed phenomenon known as photosynthesis to appear in all basic biology textbooks regardless of cultural location. However, science is far more than a distilled and purified set of objective facts that compel acceptance. It is no longer tenable for a teacher to claim that he or she is teaching *only* science. It makes sense that an isolated scientific concept (e.g., photosynthesis) is *a*cultural, but not the milieu in which the teacher, textbook, or curriculum situates the concept. About textbooks, the American critical theorist Michael Apple wrote, "They signify - through their content *and* form - particular constructions of reality, particular ways of selecting and organizing that vast universe of possible knowledge. They embody... [a] *selective tradition* - someone's selection, someone's vision of legitimate knowledge and culture, one that in the process of enfranchising one group's cultural capital disenfranchises another's" (1992, p. 5). The degree of similarity among science textbooks and curricula across cultures is unwarranted and counter-productive from a social constructivist perspective.

Assumptions about knowledge and reality, values and purpose, people and society that under gird *modern* science are grounded in Western materialism. This point has been thoroughly addressed by those interested in feminism, culture, and religion. However, curriculum adaptation too often has failed to address underlying

---

[6] In addition to the Gross & Levitt (1993) book, the current literature on anti-science includes Dyson (1993), Finneran (1996), Holton (1993), *and Scientific American* (1997). For European concerns see *Nature* (1997). For concerns in India see Nanda (1997)

assumptions primarily because the policy makers involved have tacitly accepted a cultural deficit and rationality gap viewpoint. The failure to recognize the need for authentic cultural sensitivity with regard to these assumptions has led Third World science education into social difficulties. In a series of studies between 1972 and 1980, Maddock (1983) found that science education in Papua New Guinea had a significant alienating effect that separated students from their traditional culture, "...the more formal schooling a person had received, the greater the alienation..." (p. 32). Several researchers have had similar findings (e.g., Holmes, 1977; Wilson, 1981). So, whose interest is being served?

The good of any nation or society involves several, often competing, interests. The good is rarely based on a single issue even one as important as the advancement of scientific learning. The advancement of science and science education often competes with national interest in maintaining the integrity of traditional culture. It is thus necessary to ask about the balance between these two interests, science and culture. One should ask to what extent efforts to promote scientific literacy in non-Western nations have *inadvertently* and *unnecessarily* promoted a Western, or otherwise alien, worldview? One must take to task the implicit assumption in much of the literature that non-Western, non-scientific ideas are inherently irrational, an assumption grounded in the positivist ideology that scientific thinking is the ultimate measure of rationality. People do not believe things that do not make sense. They believe precisely because sense *is* being made – because there is rationality. A reader would be mistaken to infer that this discussion is soft on superstition, or that science is being reduced to an aspect of cultural relativism. The philosopher Michael Matthews (1992, p. 14) has commented that, "many ideas and constellations of ideas can 'make sense' for an individual, and this has absolutely no connection with their truthfulness or legitimacy. Constructivists reasonably point out that teaching truths that simply do not make sense to individuals, and that forever remain foreign to them, is hardly good pedagogy; but there is a middle ground between this and endorsing... the claim that just making sense is the goal of education." Indeed, *there is* middle ground to be discovered but it will not be discovered if the focus of attention is always on the matter of traditional culture and its potentially *adverse* influence on science education - which is all too common among science educators (e.g., Gallagher & Dawson, 1986).

A Different Set of Questions

Earlier I referred to an African study in which the authors concluded that the rural people taking part in the research were irrational because they used traditional ideas to explain phenomena in nature. The renowned anthropologist Franz Boas would have had a very different view. "(T)he traditional and customary beliefs of a society provide no evidence about the way individuals think. Beliefs that an outsider considers bizarre are not evidence of bizarre thinking. They tell us something about the social tradition... about patterns of thought, which are a social product" (quoted in Musgrove, 1982, p. 70). Robin Horton (1967a; 1967b) and Yehuda Elkana (1977) concur. They argue persuasively that the cognitive activity of traditional cultures is far from primitive though clearly not scientific in the modern Western sense. Jean Lave's studies of mathematical problem solving among Africans involved with traditional trades empirically corroborate the Horton/Elkana thesis (Lave, 1988) as does other research on everyday cognition. Traditional culture poses no threat to logic and thus on these grounds need not be view as an impediment to the learning of modern science. In contrast, studies that tacitly suggest that culture is a threat make no attempt to understand how a purportedly superstitious explanation of a phenomena might be eminently reasonable from within the person's indigenous culture. Logical thinking in this research, and many others, assumes a Western-based understanding of phenomena. Clearly, anyone who maintains a culturally specific view of the world is not going to score well on these measures of logic. This is unfortunate because the promotion of science learning does not require a focus on logical thinking, but a focus on understanding.

Science content is science content regardless of culture but not so with its communication. Communicated science, which includes science education, is embedded in culture. In the jargon of education, there is always a *hidden curriculum*. This raises two issues that have received little attention. First is the issue of a potentially adverse influence of an alien hidden curriculum on the integrity of a traditional culture. The second issue concerns the potentially adverse influence on science education among those who are alienated by an alien hidden curriculum. We may not understand the complexities of culture change and adaptation, but culture does change. Any new idea brings change as people in the host environment react and adapt to the new idea. Modern science will influence a Non-western culture as surely as it has influenced Western culture. The proper concern is not about cultural change per se, but about unwarranted influence and hidden values.

1) Must all cultures adapt to science and adapt science to local culture in exactly the same way?

2) To what extent can science be taught without the cultural dress of modernism?

3) To what extent does the garb of modernism inhibit the learning of scientific concepts?

4) What cultural changes are necessary for effective science learning and what changes are unnecessary?

The implication here is that science education everywhere is too much alike and too much in the form of Western modernism. Of course, if one looks at equipment and materials, facilities, and teacher preparation there is tremendous variation. The similarities lie with the political, philosophical, and curricula of science education; and then these worldwide tend to reflect what critics call the *mythology of school science* (e.g., Smolicz & Nunan, 1975).[7]

The Culture of School Science

There are a number of scholars in science education who are strongly critical of the form and direction science education has taken in the USA and other Western nations. Too often it is scientistic and promotes a culture of school science or

[7] Milne & Taylor address the various myths of science and school science in Chapter 2.

worldview that needlessly alienates many students. Figure 4 shows seven categories that compose a worldview (see Cobern, 1991; 1993) The descriptors in the center column come from research that critically examined the cultural form in which Western science is embedded, and is employed here only as an example (e.g., Capra, 1982; Merchant, 1989, Skolimowski, 1988; Whatley, 1989). Nevertheless, there is considerable research that suggests this is a relatively accurate description of

Figure 4-Example Descriptors for Worldview Categories

Worldview Categories	Scientistic Descriptors	Alternative Descriptors
The Other or NonSelf	materialistic reductionistic exploitive	holistic social/humanistic aesthetic religious
Classification	natural only	natural social supernatural
Causality	universal mechanistic structure/functional	context bound mystical teleological
Relationship	strict objectivism nonpersonal	subjective personal
Self	dispassionate independent logical	passionate dependent intuitive
Time & Space	abstract formalism	participatory-medium tangible

Western science education. Thus, it can be argued, for example, that the scientific view of the world (i.e., all that is *Other* than one's *Self*) as presented in the classroom is often materialistic, reductionistic, and exploitive. In contrast, students may bring a holistic view of the world with a focus on social and humanistic aspects of the world. Or, the scientist of the classroom is the stereotypical dispassionate, objectively rational man. Some students, on the other hand, may be people who are quite passionate and who blend rationality, emotion, and intuition- and they may not be male. This scientistic worldview is grounded in the three imperatives of modern Western society:

The Imperative of *Naturalism* - All phenomena can ultimately and adequately be understood in naturalistic terms.

The *Scientistic* Imperative - Anything that can be studied, should be studied.

The *Technocratic* Imperative - Any device that can be made, should be made.

These imperatives serve a single goal: the material well being of people. America's most prestigious scientific organization, the National Academy of Science put it this way. "In a nation whose people depend on scientific progress for their health, economic gains, and national security, it is of utmost importance that our students understand science as a system of study, so that by building on past achievements they can maintain the pace of scientific progress and ensure the continued emergence of results that can benefit mankind" (NAS, 1984, p. 6). This is strategic thinking and it is found in UNESCO reports, the *International Journal of Science Education*, ICASE's *Science Education International*, etc.

Two points need to be made about this similarity. First, science is a necessary component of advanced technological development. Technology is a necessary component of economic development – necessary but not sufficient.[8] There are many other factors including factors of justice and morality and values that influence all facets of development. Any society, including Western societies, that relies solely on science for the development of technology and economics will be badly disappointed. In other words, however important science and technology are for economic development, they are not in themselves sufficient for economic development. Second, there is more to life than economics. In the words of that Russian Jeremiah, Aleksandr Solzhenitsyn[9] (1995, p. 8-9), modern

> culture... grows poorer and dimmer, no matter how it tries to drown out its decline by the din of empty novelties. As creature comforts continue to improve for the average person, so spiritual development grows stagnant. Surfeit brings with it a nagging sadness of heart, as we sense that the whirlpool of pleasures does not bring satisfaction, and that before long, it may suffocate us.... No, all hope cannot be pinned on science, technology, economic growth.

Jesus of Nazareth said it most eloquently, *Man does not live by bread alone*. The economic factor in science is not sufficient to maintain interest in science given the incessant reductionistic pressure of the three imperatives of naturalism, scientism, and technicism which wear away at our views of reality that give meaning to life. The principle problem in American science education today is not lower test scores than the Japanese. It is the loss of meaning. The principle problem in American science education today is the loss of meaning – and economic gain at the cost of meaningful life is a Faustian bargain.

Now let me connect these last two remarks with my earlier comments. Accepting the tight, linear science-technology-economic development (STD) model squeezes out non-scientific ways of knowing and in doing so creates for science (in

[8] Drori in Chapter 3 addresses this point at length.

[9] Two other Eastern Europeans, Polish-born poet Czeslaw Milosz (1995) and Chech scholar-statesman Vaclav Havel (1994), have addressed the corrosive affect of unbridled science and technology on modern culture.

its scientistic form) a privileged status in society. As this occurs there is increasing pressure for other aspects of culture to conform to scientific thinking. Any areas of resistance come to be viewed as deficiencies because the areas of resistance impede the takeover by scientific rationality. In other words, once the tight linear STD model is accepted, it is very difficult to avoid the charges of cultural deficit and rationality gap. This strategy may in the short run bring some economic growth – but it will do so at great human cost. It may well be that the wisest thing any group of science educators can do is to rethink the strategic issues concerning science education. What are we thinking that science can do for us? Have we like the National Academy of Science made a virtual god of science – the one on whom we depend for our well being? In our strategy for science education, what role have we given to the humanities, the arts, and to religion and morality and values? Or have we accepted the scientistic view that these are all personal and subjective where science is public and objective? One's choice of answers for these questions is critical to the form strategic science education will take and whether there will be authentic alternative constructions of science education.

References

Abiola, E. T. (1971). Understanding the African school child. *West African Journal of Education, 15*(1), 63-67.

Adas, M. (1989). *Machines as the measure of man: Science, technology, and ideologies of western dominance*. Ithaca, NY: Cornell University Press.

Aikenhead, G. S. (1996). Science education: Border crossing into the subculture of science. *Studies in Science Education, 27*, 1-52.

Altbach, P. G. (1987). *The knowledge context: Comparative perspectives on the distribution of knowledge*. Albany, NY: SUNY Press.

Apple, M. W. (1992). The text and cultural politics. *Educational Researcher, 21*(7), 4-11, 19.

Basalla, G. (1967). The spread of western science. *Science, 156*, 611-622.

Capra, F. (1982). *The turning point: Science, society, and the rising culture*. New York, NY: Simon and Schuster.

Champagne, D. W., & Saltman, M. A. (1964). Science curricula and the needs of Africa. *West African Journal of Education, 8*(3), 148-150.

Cobern, W. W. (1991). *World view theory and science education research*, NARST Monograph No. 3. Manhattan, KS: National Association for Research in Science Teaching.

Cobern, W. W. (1993). Contextual constructivism: The impact of culture on the learning and teaching of science. In K. G. Tobin (editor), *The practice of constructivism in science education* (pp. 51-69). Hillsdale, NJ: Lawrence Erlbaum Associates, Inc.

Cobern, W. W. (1994). Point: Belief, understanding, and the teaching of evolution. *Journal of Research in Science Teaching, 31* (5), 583-590.

Cobern, W. W. (1996a). Constructivism and non-Western science education research. *International Journal of Science Education, 18*(3), 295-310.

Cobern, W. W. (1996b). Worldview theory and conceptual change in science education. *Science Education, 80*(5), 579-610.

Cobern, W. W. & Aikenhead, G. (1997). Culture and the learning of science. In B. Fraser, & K. G. Tobin (editors), *The international handbook on science education*. Dortrecht: Kluwer Academic Publishers.

Dart, F. E., & Pradhan, P. L. (1967). Cross-cultural teaching of science. *Science, 155*(3763), 649-656.

Dedijer, S. (1962). Measuring the growth of science. *Science, 138*(3542), 781-788.

Dyson, F. J. (1993). Science in trouble. *American Scholar, 62*(4), 513-525.

Elkana, Y. (1977). The distinctiveness and universality of science: reflections on the work of Professor Robin Horton. *Minerva, 15*, 155-173.

Finneran, K. (1996). Can science get any respect? *Issues in Science and Technology, 13*(1), 95-96.

Gallagher, J. J. & Dawson, G. (1986). *Science education & cultural environments in the Americas: A report of the Inter-American seminar on science education*. Washington, DC: The National Science Teachers Association.

Garrison, J. W., & Bentley, M. L. (1990). Science education, conceptual change and breaking with everyday experience. *Studies in Philosophy and Education, 10*(1), 19-35.

Geertz, C. (1973). *The interpretation of culture*. New York, NY: Basic Books.

Gross, P., & Levitt, N. (1993). *Higher superstition: The academic left and its quarrels with science*. Baltimore, MD: John Hopkins University Press.

Hills, G. L. C. (1989). Students' "untutored" beliefs about natural phenomena: primitive science or commonsense? *Science Education, 73*(2), 155-186.

Holmes, B. (1977). Science education: Cultural borrowing and comparative research. *Studies in Science Education, 4*, 83-110.

Holton, G. (1993). *Science and anti-science*. Cambridge, MA: Harvard University Press.

Horton, R. (1967). African traditional thought and Western science, Part I. From tradition to science. *Africa, 37*, 50-71.

Horton, R. (1967). African traditional thought and Western science, Part II. The 'closed' and 'open' predicaments. *Africa, 37*, 155-187.

Lave, J. (1988). Cognitive consequences of traditional apprenticeship training in West Africa. *Anthropology & Education Quarterly, VIII*(3), 177-180.

Lawson, A. E., & Weser, J. (1990). The Rejection of Nonscientific Beliefs about Life: Effects of Instruction and Reasoning Skills. *Journal of Research in Science Teaching, 27*(6), 589-606.

Lewin, K. M. (1990). International perspectives on the development of science education: Food for thought. *Studies in Science Education, 18*, 1-23.

Lewin, K. M. (1993). Planning policy on science education in developing countries. *International Journal of Science Education, 15*(1), 1-15.

Lord, J. (1958). The impact of education on non-scientific beliefs in Ethiopia. *Journal of Social Psychology, 47*, 339-353.

Maddock, M. N. (1981). Formal schooling and the attitudes of Papua New Guinean students 1972-1980. *Research in Science Education, 11*, 180-192.

Maddock, M. N. (1983). Research into attitudes and the science curriculum in Papua New Guinea. *Journal of Science and Mathematics Education in S.E. Asia, VI*(1), 23-35.

Matthews, M. R. (1992). Old wine in new bottles: a problem with constructivist epistemology. Paper presented at the annual meeting of the *National Association for Research in Science Teaching*.

Merchant, C. (1989). *The death of nature: Women, ecology, and the scientific revolution*. San Francisco, CA: Harper & Row.

Musgrove, F. (1982). *Education and anthropology: Other cultures and the teacher*. New York, NY: John Wiley & Sons.

Nanda, M. (1997). The Science wars in India. *Dissent, 44*(1). <http://www.igc.apc.org/dissent/archive /winter97/nanda.html>

National Academy of Sciences. (1984). *Science and creationism: a view from the National Academy of Sciences*. Washington, DC: National Academy of Sciences.

Nature. (1997). The 'Sokal affair' takes transatlantic turn. *Nature, 385*(6615), 381.

Okebukola, P. A., & Jegede, O. J. (1990). Eco-cultural influences upon students' concept attainment in science. *Journal of Research in Science Teaching, 27*(7), 661-669.

Poole, H. E. (1968). The effect of urbanization upon scientific concept attainment among Hausa children of Northern Nigeria. *British Journal of Educational Psychology, 38*, 57-63.

Prather, J. P. (1990). *Tracing science teaching*. Washington, DC: National Science Teachers Association.

Rostow, W. W. (1971). *The stages of economic growth: a non-communist manifesto*. Cambridge, UK: Cambridge University Press.

Salam, A. (1984). *Ideals and realities: Selected essays of Abdus Salam*. Singapore: World Scientific.

Scientific American. (1997). Science versus antiscience? *Scientific American, 276*(1), 96-101.

Shapiro, R. (1986). *Origins: a skeptic's guide to the creation of life on earth*. New York, NY: Summit Books.

Skolimowski, H. (1988). Eco-philosophy and deep ecology. *The Ecologist, 18*(4/5), 124-127.

Smedslund, J. (1970). Circular relation between understanding and logic. *Scandinavian Journal of Psychology, II*, 217-219.

Snow, C. P. (1961). The moral un-neutrality of science. *Science, 133*(3448), 255-262.

Urevbu, A. O. (1987). School science in West Africa: An assessment of the pedagogic impact of Third World investment. *International Journal of Science Education, 9*(1), 3-12.
Whatley, M. H. (1989). A feeling for science: Female students and biology texts. *Women's Studies International Forum, 12*(3), 355-362.
Wilson, B. (1981). The cultural contexts of science and mathematics education: Preparation of a bibliographic guide. *Studies in Science Education, 8*, 27-44.

Cathrine E. Milne & Peter C. Taylor

Chapter 2

Between a Myth and a Hard Place: Situating School Science in a Climate of Critical Cultural Reform

These are the days that must lay a new foundation of a more magnificent philosophy, never to be overthrown, that will Empirically and Sensibly canvass the Phenomena of Nature, deducing the causes of things from such Originals in Nature as we observe are producible by Art, and the infallible demonstration of Mechanicks: and certainly this is the way, and no other to build a true and permanent Philosophy.... [T]o find the various turnings and mysterious process of this divine Art, in the management of this great Machine of the World, must needs be the proper Office of only the Experimental and Mechanical Philosopher.

(Henry Power, 1664)[10]

Our experiences as teachers and educational researchers indicate a general enthusiasm amongst science teachers for constructivist-based teaching practices aimed at improving the quality of student learning. However, even when teachers believe that constructivism is an appropriate epistemology (or way of knowing), they struggle to implement and maintain teaching practices informed by constructivist theory (Taylor, 1996; Tobin, Davis, Shaw & Jakubowski, 1991; Vance & Miller, 1995). We believe that the difficulties experienced by science teachers in instituting constructivist-inspired changes in their classrooms can be explained, in large part, if school science is viewed as a cultural activity which is constrained by powerful and ubiquitous cultural myths.

The teaching of Western school science is a cultural activity inasmuch as it involves teachers engaging students in discursive practices (or discussion-centred activities such as science laboratory investigations) designed to enable them to understand something about that highly-valued aspect of their cultural heritage known as 'Science'. Our interest in exploring the nature of the discursive practices of the culture of school science arises from research that shows that teachers are largely unaware of the mythical status of many school science practices. Our concern is to

---

10 *Experimental philosophy, in three books, containing new experiments: Microscopical, mercurial, magnetical: With some deductions, and probable hypotheses raised from them, in avouchment and illustration of the now famous atomical hypothesis*, **Johnson Reprint, New York. (reprint of the 1664 edition, with the addition of Power's notes, corrections, and emendations, and a new introduction by Marie Boas Hall, 1966).**

W. W. Cobern (ed.), Socio-Cultural Perspectives on Science Education, 25–48.

draw attention to the subsequently impoverished and disempowering image of the nature of science and scientific knowledge that teachers and students co-construct.

The role of myths – *narrative accounts of collective experience* – in shaping the historical development of cultural groups has been the subject of extensive research by anthropologists,[11] sociologists,[12] psychologists,[13] socio-linguists,[14] historians[15] and semioticians.[16] Although far from united in their views on the nature of myth, scholars have highlighted the central role of myth in legitimating particular ways of knowing, valuing, feeling, and acting that regulate and order social reality:

> Myth expresses, enhances and codifies belief; it safeguards and enforces morality; it vouchsafes for the efficiency of ritual and contains practical rules for the guidance of man. Myth is thus a vital ingredient of human civilization; it is not an idle role but a hard worked active force. (Malinowski, 1971, p. 19)

However, the power of myth in creating social coherence exacts a cost. While supplying a cultural group with a natural image of reality, its presence in the discursive practices that it shapes, and through which it is propagated, is masked and, in time, forgotten. Consequently, the historical and contingent quality of established patterns of beliefs and practices is replaced by an unwarranted sense of naturalness and inevitability. "Myth is constituted by the loss of the historical quality of things: in it, things lose the memory that they once were made" (Barthes, 1972, p. 142). Constructivist-oriented philosophers of science describe science as a distinct culture composed of its own socially constructed notions (Kuhn, 1970; Toulmin, 1990). This leads us to argue that, like any other culture, myths emerge in science as a cultural force and serve to propagate universal ideas about the practice of science and the legitimation of scientific knowledge.

In the field of education, Britzmann (1991) alerts us, from a poststructuralist perspective, to the dangers of repressive myths (e.g., *teachers are self-made, experience makes the teacher*) that mask powerful forces of enculturation governing the preparation of student teachers for the profession of teaching. In this chapter, we focus our inquiry on myths of science education, especially disempowering pedagogical myths that have their genesis in the 17th century development of experimental science. We argue that teachers and students (and curriculum developers, textbook writers, examiners) accept these myths as the natural and inevitable social reality of science and the science classroom. The historical memory loss accompanying these myths is a characteristic of contemporary school science and is instrumental in neutralising the power of constructivist-based teaching reforms.

Although we value the trend to constructivist transformation of school science, we believe that, because contemporary constructivist perspectives tend to exclude considerations of the historical quality of science, especially experimental science,

[11] See for example Douglas (1962), Geertz (1973), Levi-Strauss (1970).
[12] See for example Durkheim (1976), Malinowski (1954, 1971).
[13] See for example Freud (1965), Jung (1968).
[14] See for example Lakoff & Johnson (1980).
[15] See for example Campbell (1968), Slotkin (1973).
[16] See for example Barthes (1972).

long-established myths about the nature of science are likely to continue to serve as legitimating frameworks for the culture of classroom life.

Indeed, we believe that to teach science without reference to myths is much harder than to teach within a framework of myths about science. Smolicz and Nunan (1975) describe myths as "irreducible assumptions". We interpret this definition of myth to imply that a believer of a myth related to science cannot imagine an alternative perspective about science. Our attempt to identify myths of school science led us to examine the literature on 17th century European experimental philosophers. We identified the following questions that fuelled crucial debates whose outcomes have served to legitimate the discursive practices of Western experimental science. We argue that echoes of these seventeenth century debates continue to resonate strongly in school science classrooms.

- Which is eternal – objects, words or ideas?
- How is scientific knowledge generated?
- How should we (re)present this knowledge?

However, we did not conduct our analysis from a neutral perspective. As proponents of constructivist theory interested in enabling teachers to deconstruct epistemological barriers to constructivist-related teaching reform, our reading of history was interpreted from a constructivist perspective enriched by critical theory. A major aspect of the program of critical social theory is to challenge the 'taken-for-granted' assumptions that constitute social reality and social structures as homogeneous, unproblematic and predictable. In this paper, we are trying to help teachers to appreciate the central role they have in constructing, maintaining and legitimating (usually unwittingly) these social structures by implementing a pedagogy that is underpinned by science cultural myths.

THE REFERENT OF CRITICAL CONSTRUCTIVISM

Critical constructivism is an epistemology that can serve as a referent for cultural reform in science education because it foregrounds the normally invisible socio-cultural context of knowledge construction. Drawing on the critical theory of Jurgen Habermas, critical constructivism challenges objectivism, the view that absolute knowledge is available to science, and unmasks the subtlety of its iron grip on official (i.e., institutional, bureaucratic) ways of knowing. For science education, critical constructivism provides a powerful theoretical framework for under-standing the contemporary culture of school science, for making visible repressive cultural myths that govern social roles and practices, and for considering what constitutes appropriate communicative relationships amongst teachers and students (Taylor, 1996, in press).

Based on Habermas's (1972) theory of knowledge and human interests, there are three fundamental ways of knowing – *technical, communicative, critical* – which have their roots in the natural history of the human species. Each way of knowing arises from a distinctive mode of goal-oriented reasoning and promotes a worldview that serves particular interests or goals (McCarthy, 1978; Pusey, 1987).

The first, *technical rationality*, fuels the self-interest goal of species survival and reproduction, and is manifested in a desire to control and exploit the environment. One of the global consequences of an unbalanced technical rationality is the detrimental effect of our attempts to conquer nature by perpetuating the myth of Nature as an exploitable resource (Bowers, 1987). How did technical rationality gain its official status? Schon (1983) explains that technical rationality underpins the *myth of positivism* that enshrined the empirical-analytical sciences as the source of secure and privileged knowledge of the world. Technical rationality gained an exalted status in the Enlightenment and colonised the social sciences and, later, education as they were admitted to the university.

During the twentieth century, the dominance of technical rationality in the industrialised Western world, especially in its social institutions, has resulted in an unbalanced way of knowing and acting, and a crisis of confidence within many of the major professions. Today, technical rationality is manifested in economic rationalist goals of efficiency and productivity, goals that threaten to displace a concern with people as ends in themselves. It is our intention to reveal the embedded presence in school science of the implicit objectivism of technical rationality, thereby providing a basis for science educators to work towards deconstructing the cultural myths of *cold reason* and *hard control* which act in concert to preserve science curricula from constructivist-inspired epistemological reform (Taylor, in press, 1996).

We are able to pursue this goal because of Habermas's rejection of the "objectivist illusion" of the technical view of science "as a universe of facts independent of the knower, whose task it is to describe them as they are in themselves" (McCarthy, 1985, p. 59). Habermas identifies the central role of subjectivity in creating and validating knowledge, especially knowledge derived by empirical-analytical means (including science and mathe-matics). In a Kantian sense, "knowledge is necessarily defined both by the objects of experience and by a priori categories and concepts that the knowing subject brings to every act of thought and perception" (Pusey, 1987, p. 22). This view is in close accord with the radical constructivist position of Ernst von Glasersfeld (1990, 1991) who draws on Piaget's genetic epistemology to argue that cognition is a dynamic self-organisation of experiential reality, rather than an act of discovery of an objective ontological reality (Hardy & Taylor, 1997).

However, Habermas departs from the implicit individualism of Kant and focuses on the socio-cultural and historical contexts of knowledge construction. From his later theory of communicative action (Habermas, 1984), which aims to show that "there is no knower without culture and that all knowledge is mediated by social experience (Pusey, 1987, p. 23), arises a second way of knowing based on a *communicative rationality*. The central goal of this way of knowing is to communicate with others for the purpose of achieving mutual and reciprocal understanding. Notwith-standing the pluralistic nature of Western society and the multitude of competing interests, society depends for its cohesion and integrity on the central goal of communicative rationality.

In cultural anthropology, communicative rationality is embedded in interpretive (or hermeneutic) modes of ethnographic inquiry that seek to understand the meaning-perspectives of actors in social situations (Erickson, 1986). However, a limitation of interpretive understanding is that it fails to critique the seemingly natural and self-sufficient cultural frameworks that shape (or distort) participants' meaning-perspectives. "Hermeneutics comes up against walls of the traditional framework from the inside, as it were" (Habermas, 1978, p. 44). Thus, a third way of knowing, involving *critical rationality*, arises from a concern with organising social relations on the basis of communication that is free from the distorting influence of ideologically-oriented interests associated with actions arising from an unchecked (and invisible) technical rationality. Critical rationality promotes the emancipatory goal of self-critical reflective knowledge, but within a critical theory of social action that aims to reform society rather than only understand it. In school science, therefore, an important emphasis is on raising students' critical awareness of socio-cultural myths, such as objectivism, that work in concert with other technical myths to beguile (teachers and) students into accepting a disempowering sense of a lack of agency (or self-determination) as learners.

In summary, critical constructivism embraces communicative and critical ways of knowing, and offers a powerful referent for the transformation of school science discursive practices that have been governed traditionally by a pervasive technical rationality. Because of its explicit axiological commitment, critical constructivism offers ethical principles for establishing critical communicative relationships (Taylor, in press; Vandenberg, 1990). In school science, educative relationships between teachers and students should be sustained by a commitment to achieving: (1) equality of opportunity for enacting dialogical[17] roles, rather than domination by more powerful participants (i.e., the teacher, high-achieving students, disruptive/unruly students); and (2) critical awareness of the culturally-framed (often invisible) norms of rational discourse, rather than naive acceptance as tends to be the case in everyday communicative interactions. In the science class of the future, we envisage open and critical discourses that involve interpretive inquiry, critique of ideology, analysis of social systems, and a philosophy of history (Taylor & Williams, 1993).

Our purpose in writing this chapter is, therefore, to enable science educators to work towards creating empowering educative relationships amongst teachers and students. Clearly, an important first step is for teachers to become self-critically reflective about the historically situated nature of the extant discursive practices that they promote in their science classrooms. Therefore, the central question of this chapter is: What unseen cultural myths associated historically with the development of science exert a major influence on the discursive practices of the contemporary school science classroom? In attempting to find an answer to this question, our critical constructivist perspective served as an epistemological lens through which to reflect critically on the implications of history for the discursive practices of contemporary school science classrooms. Our specific focus is on science practical activities.

---

[17] A dialogical discourse is communication that aims to create a rich mutual understanding of each other's point of view or valued beliefs without either party feeling obliged to adopt the other's position simply because of a power imbalance in the relationship. Indeed, because a dialogical understanding can embrace opposing or conflicting beliefs, it serves to enhance the 'less powerful' participant's sense of agency as a rational thinker, actor, and communicator (Bakhtin, 1981).

MYTHS AND SCHOOL SCIENCE

During the 17th century, European experimental philosophers were engaged in debating several major questions whose outcomes served to legitimate the discursive practices of Western experimental science. In revisiting these debates, we have identified a number of key cultural issues that, over time, evolved from belief into myth. We believe that these science cultural myths are instrumental in governing the epistemology of contemporary school science, particularly practical activities conducted in school science laboratories. By making visible these myths we can recognise the historical contingency of a number of discursive practices of contemporary school science that have become so routinised that they have taken on an aura of naturalness and inevitability. We feel that it is important to understand how these myths, if left unexamined or uncontested, are likely to maintain a culture of school science that is resilient to constructivist-related curricula reforms.

Question 1: Which is Eternal: Objects, Words or Ideas?

Science has its antecedents in the writings and thoughts of Aristotle, as interpreted by ancient and Medieval Greek and Arab scholars and in the development of experimental science in Europe from the 17th century onwards. Like Plato, his teacher, Aristotle argued for the existence of universals (or eternal and immutable ideas). However, unlike Plato, who believed that reason could generate knowledge of universals that existed in a separate world of ideas, Aristotle believed that direct examination of the matter and form of objects provided evidence about the nature of universals. Whereas Plato assigned a privileged status to reason and the world of ideas, for Aristotle knowledge of universals was based on sensory evidence of objects, particularly living things, as observed directly in nature (Gaarder, 1995). In 17th century Europe, a debate flourished about whether universals were constructions of the mind or whether they existed externally as part of reality (Slaughter, 1982). Advocates of these disparate perspectives contested the ontological question of the permanency of objects or ideas.

The French natural philosopher, René Descartes, proposed a philosophy of *idealism* which posited that knowledge of objects is formed in the mind, independently of the senses (Descartes, 1642). It was only in relation to the existence of God that Descartes could develop a separate argument to support his *idealistic* notion that human knowledge of the external world comes through the mind rather than through the senses. Descartes argued that his idealism required the existence of a God because His role was to be the final arbiter of the veracity of human thought. Descartes' *rationalistic* philosophy made human reason pre-eminent to the observation of nature. Thus, new knowledge could be known only via the mind because the mind, unlike the senses, was immune to being misled.

Natural philosophers who believed in a direct correspondence between the notion of an object and the object itself came to be called realists. For most 17th century experimental philosophers, objects were permanent and created a reality external to humans. An understanding of the nature of universals could be achieved, therefore, by examining Nature. Whereas idealists such as Descartes supported the existence of God as the final arbiter of ideas, realists of the early 17th century believed that observations of the natural world provided a greater understanding of

God's creation. Careful use of their senses placed them in direct communication with the natural world and gave a truer picture of God's creation than ever could be achieved by use of words and ideas. Relative to human existence, external reality was fixed and behaved in a consistent way that allowed experimental philosophers to assume that experiments could be conducted over time and that the data collected as a result represented an unchanging reality. From this realist perspective, a direct relationship between a phenomenon in Nature and the word used to name the phenomenon was accepted (Slaughter, 1982). According to realists, because God created the world She/He created a finite number of objects that humans could learn about by examining Nature. This concept led to the realist myth of a 'tree-like structure' of knowledge: knowledge is based on a finite number of objects that exist in the universe, and observations of these worldly objects represent facts about the natural world (Arbib & Hesse, 1986). This belief that by observations one could experience reality led to the emergence of the myth that human observations mirrored Nature's reality.

> Science Cultural Myth of Naive Realism:
> Our observations of reality correspond exactly
> to an external reality.

The naive realist perspective won the 17th century debate on universals, probably because it was more congruent with experimental philosophers' (culturally-framed) aspirations, common sense understandings of their experiences, and their belief in the direct relationship between Nature, God and their senses. Thus, naive realism adopted the role of a powerful cultural myth that became entrenched by 'disappearing from view'. Over time, perceptions waned of the historical properties of naive realism and its relationship with a religious worldview as the consciousness of its proponents became saturated by an overwhelming sense of the naturalness of realism. Its apparent naturalness masked its historio-cultural contingency, as it became THE guiding framework for the conduct and legitimation of experimental science. Amongst its proponents, this mythical framework legitimated discursive practices that were believed to yield privileged knowledge of the laws of Nature. However, with the evolution of scientific understanding came the realisation that perhaps we could not observe reality exactly. Nevertheless in school science, naive realism, as proposed by 17th century realists, continues to operate as a powerful myth.

Realism and Pedagogy: In contemporary school science, where realist myths remain unrecognised and uncontested, an illusion of the certainty of knowledge generated is woven into the teaching and learning of science. Within this illusion a seamless tapestry of *a*historic scientific facts materialises before students' eyes. The disempowering spell of the myth of realism is wholly captivating when students believe that they can see scientific facts by looking ever outwards (at Nature, the textbook, the blackboard, the teacher, the experimental equipment) rather than inwards (at their own conceptions).

As a psychological theory of learning, constructivism urges science teachers to focus students' attention on their own conceptual frameworks, and to use Nature as a testing ground for the viability of their conceptions, both old and new. However, the moral certitude of a God's-eye view of Nature (through the microscope, telescope or test tube) is a formidable guardian of naive realism and the notion that knowledge can exist in objects that are external to the learner. Consequently, the notion that humans could be involved in the construction of observations of Nature can be ignored. However, if constructivism is to challenge the perceived relationship between Nature, observation and God then it must go beyond a psychology of learning and be transformed into a philosophy of learning, one that can contest philosophically inspired cultural myths.

The emergence of 'radical' and 'critical' forms of constructivism is an attempt to achieve this transformation. Radical constructivism's *experiential realism* (science is a relative truth) contests the myths of *naive realism* (science is the absolute truth) and *scientific realism* (science is approximating the absolute truth) by construing the learner (student, teacher, or scientist) as a knowing being concerned with making good sense of the world (Von Glasersfeld, 1990, 1991). Competing knowledge claims are judged not in terms of their accuracy in mirroring Nature, but in terms of their viability for achieving the learner's goals. Radical constructivism rejects the mantle of idealism, particularly its mutant variety, solipsism (i.e., the world is as I dream it to be), by acknowledging openly the existence of objective reality, particularly its constraining influence on the learner's attempts to make good sense of the world. Critical constructivism's concern with a moral basis for judging the social worth of competing knowledge claims rescues constructivism from the spectre of moral relativism. The learner is construed as not only a knowing being but also an historio-cultural being whose communicative relations with teachers are at risk of being systematically distorted by unrecognised and uncontested cultural myths (Taylor, 1996, in press).

In order for constructivism to be successful in contesting the cultural myth of naive realism, it is important to be aware of the extent of naive realism's powerful invisible grip, both in science and school science. The power of this myth is evident in its influence on the historical resolution of epistemological questions about the nature of scientific methodology, scientific language, and genre of experimental report writing. We discuss these questions in the following sections.

Question 2: How is Scientific Knowledge Generated?

For 17th century experimental philosophers, the answer was clear: observation and experimentation generate knowledge. Francis Bacon based his proposal for experimental philosophy on the notion that worthwhile knowledge could be generated and legitimated only through the use of observation and experimentation. He promoted inductivism as the appropriate approach for the conduct of experimental philosophy because inductivism tied together the pre-eminence of sensory observations to the search for universals.

In previous ages, however, the appropriate method for legitimating knowledge was based on *dialectic syllogism,* a method of argumentative logic. Syllogistic logic used deductive reasoning to generate true conclusions from axioms which could be found in authoritative texts composed of the works of Aristotle with commentaries by Greek and Arab scholars. Dialectic reasoning was aimed at finding arguments for and against particular position statements. In 17th century Italy, legitimation of authorship and knowledge was controlled by absolute rulers so that claims were presented either namelessly in courtly fiction (Biagioli, 1992), or depended on the authentication of the absolute ruler (Findlen, 1993). The difficulty with dialectical syllogism was that it resulted in a disparate collection of ideas and questions about phenomena. Because these could not be integrated into a coherent and consistent cosmology they were not systematically examinable.

Bacon's proposed philosophy of inductivism established a unifying principle for examining Nature and provided a framework for experimental philosophers to begin to develop a consistent and coherent cosmology that was open to systematic examination. Bacon argued that one must start from the senses and particulars, progress to middle axioms, then to experimentation and, finally, to general axioms. This preferred way of reaching understanding required an emphasis on experimentation, a practice that tended to be ignored by proponents of dialectic syllogism. Bacon believed that Nature was more subtle than argument or empty dogma. If observations of Nature contradicted aspects of an axiom then the axiom should be changed, but the tendency of schools of thought based on dialectic syllogism had been to rescue and preserve the axiom by some "frivolous distinction" (Bacon, 1620/1968, p. 51). His experimental philosophy provided an acceptable rationale for experimentalists such as Boyle, Hooke and Huyugens.

The Status of Scientific Facts: Bacon's inductivist experimental philosophy involved the identification of fundamental human constructs called matters of fact (or scientific facts) which he presented as a record of observations obtained during the conduct of experiments. From his naive realist perspective, discovery in science resulted from the belief that facts existed in Nature and that scientists' observations of experimental outcomes constituted observations of Nature. In contrast to scientific theories, which were believed to be dependant on human reason and tainted by the influence of human opinion, matters of fact did not seem to depend on human thought for their existence but instead were based on observation of God's creation, Nature. Thus, matters of fact came to possess an certainty that could never be afforded to scientific theories (Shapin & Schaffer, 1985).

Laboratories were developed as appropriate public areas where like-minded experimental philosophers could work co-operatively and where matters of fact could be demonstrated. These contrasted with existing practical working areas maintained by alchemists that had been secretive and closed to the public. The laboratories of the 17th century experimental philosophers formed the precursors of modern laboratories. Presenting knowledge in a public forum helped to reinforce its matter of fact status. In his *History of the Royal Society*, Thomas Sprat relates the use of the laboratory with the establishment of a matter of fact when he writes:

> Those to whom the conduct of the *Experiment* is committed... carry the eyes and the imaginations of the whole company into the *Laboratory* with them. And after they have perform'd the *Trial* they bring all the *History* of its *process* back again to the *test*. Then comes in the second great work of the *Assembly*, which is to *judg[e]* and *resolve* upon the matter of *Fact*. (Sprat, 1667, p. 169)

Robert Boyle, an early experimental philosopher, believed strongly that knowledge generated by conducting experiments in laboratories in front of witnesses should be available to other interested parties (Shapin & Schaffer, 1985). Because he believed that observations he conducted during an experiment were records of reality and represented matters of fact, he was compelled to address the issue of how he was to convince others that he had observed Nature truly. Boyle achieved this by creating a 'visual source' through the rich descriptions of his experiments (Shapin, 1984), much like those of contemporary qualitative researchers (Guba & Lincoln, 1989). Using a literary approach, Boyle intended to convince his readers that he had conducted experiments that had yielded matters of fact. For example, Boyle richly describes the results of an experiment that support the notion that air quality is important for continued life:

> **We will add (by way of confirmation) the following experiment: in such a small receiver, as those wherein we killed divers birds, we carefully closed up one, who, though for a quarter of an hour he seemed not much prejudiced by the closeness of his prison, afterwards began to first pant very vehemently, and keep his bill very open, and then to appear very sick; and last of all, after some long and violent strainings, to cast up some little matter out of his stomach; which he did several times, till growing so sick that he staggered and gasped, as being just ready to die. We perceived, that within about three quarters of an hour from the time that he was put in, he had so sickened and tainted the air with the streams of his body, that it was become altogether unfit for the use of respiration[.] (Boyle, 1660, p. 105)**

Matters of fact were assigned a high status, particularly with regard to their intellectual and moral certainty. According to Bacon, because God created Nature, scientific observers were recording God's creation, which was immutable and permanent, unlike human opinion that was likely to distort the nature of things. Others agreed.

> **[I]n the works of *Nature* the deepest discoveries shew us the greatest Excellencies. An evident Argument that He that was the Author of all things was no other than *Omnipotent*. (Hooke, 1665, p.1)**

This belief in a natural theology persisted in presentations of observations and experiments in science well into the nineteenth century (Crossland, 1976). As a consequence, knowledge generated from experimentation and observation assumed an ethical as well as a cognitive dimension. It became a cultural myth of science that matters of fact existed independently of human reason.

Science Cultural Myth of Scientific Facts: Scientific facts do not depend on reason or opinion, have ethical as well as intellectual status, and a certainty not possessed by scientific theories.

Pedagogical Myths of Observation: We believe that this powerful cultural myth, which promotes unquestioned acceptance of (1) the primacy to human understanding of observations by unfettered minds and (2) the direct relationship between observation and experimentation, is a major force in governing discursive practices of the contemporary school science laboratory. We argue that the myth of scientific facts, which reinforces a naive realist belief also underpins an objectivist belief that in which knowledge exists externally of knowers. Several key pedagogical assumptions about learning science result from an adherence to this myth. Unless these assumptions are recognised and contested, they are likely to neutralise efforts aimed at constructivist reform of school science.

The first pedagogical assumption is related to the naive realist belief that everyone who observes a particular object or process, especially if they have an 'open mind', can make identical observations. In the school science classroom, the assumption that observations are independent of theory and that teacher and students are 'seeing the same thing' when observing a practical activity can lead to the further assumption that it is not necessary for the teacher to address explicitly the theoretical framework of a practical activity. Instead, students are assumed to build up their knowledge to be like that of the teacher by automatically making connections in their minds with observations that they have made previously.

> Pedagogical Myth 1:
> Everyone involved in a practical activity
> makes identical observations.

A second pedagogical assumption is that the practical work 'speaks for itself' and, consequently, there is no need for discussion of the significance of the outcomes of the practical activity for the generation of theory. If practical work speaks for itself then the theory should percolate in the student's mind direct from their observations. Figure 2 illustrates the percolator myth of knowledge generation that has led science educators to place a high value on the practical activity. Although we acknowledge that students can benefit from first hand experience of the laboratory methods used by scientists to legitimate scientific knowledge, they also need to be aware of the theoretical frameworks in which both they and scientists operate. After all, Robert Boyle and Isaac Newton did not blithely conduct a mass of unrelated experiments and then look for order. Their experimentation was based on, and informed by, underlying theory.

> Pedagogical Myth 2:
> During the practical activity,
> theories percolate into consciousness from
> observations.

The naive realist belief that human understanding of the natural world is based on observation rather than discourse continues as a powerful organising principle in school science. The assumption that doing and seeing are more important than thinking and writing can lead teachers to introduce a practical activity in the absence of a discussion of the activity's purpose (Sutton, 1992). In these situations, theories are not examined because teachers assume that students will build up theories in their minds if they make sufficient observations (see Figure 1). In school science, an overriding emphasis on observation and experimentation and the neglect of theorising is due to a particular reading of what represents 'true' science, where the practical activity has become 'the source' of scientific knowledge:

> **Pedagogical Myth 3:**
> **Observations obtained during the practical activity are more important than is discourse for developing a knowledge of Nature.**

In summary, 17th century realist beliefs in the privileged role of observation and experimentation for generating new scientific knowledge have become mythified within a Western scientific worldview. In the context of contemporary school science, these cultural myths can give rise to pedagogical assumptions that overlook the important role of theory in shaping students' observations and fail to acknowledge the socially constructed nature of scientific facts. Indeed, it is but a short step from being indoctrinated into this ideology to making the scientistic assumption that a Western scientific worldview yields an intellectually and morally privileged view of the cosmos.

From a critical constructivist perspective, all observation is framed by theory and all theory is subject to review in light of disconfirming observation. Thus, it is important for teachers of school science to make students aware of the interrelationship between scientific observation and theory, and of the consequent uncertainty of scientific facts that emerge from this process. In so doing, it makes good sense to engage students in reflecting critically on the masking role of these myths, on the historio-cultural contexts that gave rise to them, and on the political interests that continue to be served by their unacknowledged presence, both in school science and society in general.

According to some modernist philosophers of science (education), constructivism serves to perpetuate realist myths because of its apparent enthusiasm for the notion of making sense of the world through experience. Matthews (1993) argues that constructivism promotes a *naive empiricist* epistemology that encourages the belief that sensory perceptions alone lead to understanding. He claims that constructivists believe in a direct relationship between observations of the natural world and knowledge construction. Although it is possible that some (psychology-bound) constructivists are proponents of a simplistic role for experience in students' learning, we find that von Glaserfeld's (1990) notion of experiential reality provides a powerful view of the dialectical relationship between a learner's prior understandings, experience of the world, reflective thinking and knowledge construction. The notion of experiential reality transcends the search for a true external reality and adds a useful philosophical dimension to constructivism that enables science educators to begin to contest pedagogical assumptions underpinned by realist cultural myths. However, it is not until we add a critical social dimension to constructivism that we can posit an ethical basis for contesting the moral certitude of discourses of power and privilege that are governed by realist myths.

Figure 1. 'Percolator' myth of theory generation

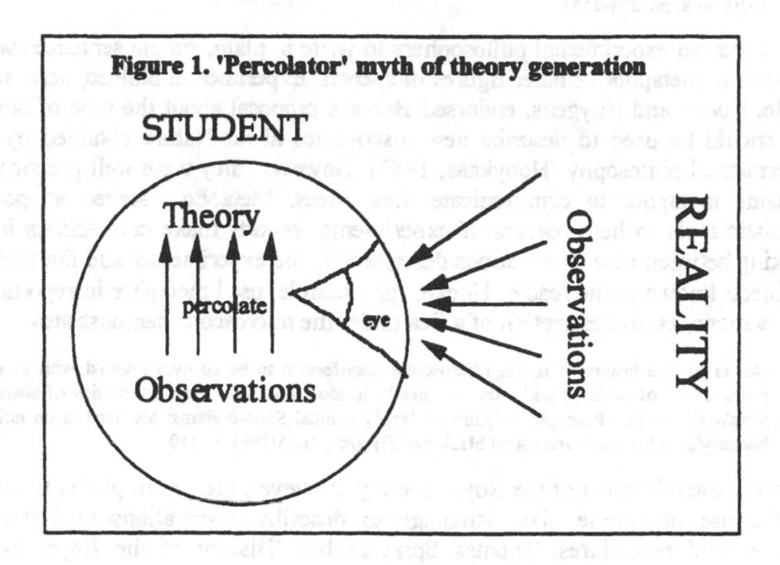

Question 3: How Should Scientific Knowledge be (Re)Presented?

This question was central to the activities of 17th century experimental philosophers who believed that their observations constituted accurate descriptions of Nature. Bacon's notion that human understanding must be based on observations of the natural world influenced both his belief in the importance of experimental philosophy and his interest in the type of language that should be used by experimental philosophers to describe their observations. Thus, scientific discourse evolved concurrently with experimental philosophy. Because it determined what counted as legitimate knowledge about the natural world, scientific discourse became the discourse of power. However, as this occurred the historical development of scientific discourse was masked and, subsequently, appeared timeless.

Against Metaphor? Bacon distinguished between prose used to present scientific observations and experiments and prose that excited the imagination (Vickers, 1985). He believed that science should report on direct examinations of the reality of the natural world and not be misled by the vagaries of verbal science that he associated with dialectic syllogism. For him, and for the scientists who followed the experimental method, language used to present observations had higher status than language used to provide verbal commentary. In his treatise, *Parasceve*, Bacon used the analogy of a person collecting materials to construct a ship to describe the use of language by scientists to construct new knowledge:

> And for all that concerns ornaments of speech, similitudes, treasury of eloquence, and such like emptiness, let it be utterly dismissed. Also let all those things which are admitted be themselves set

> down briefly and concisely, so that they may be nothing less than words. For no man who is collecting and storing up materials for shipbuilding or the like, thinks of arranging them elegantly, as in a shop, and displaying them so as to please the eye; all his care is that they be sound and good, and that they be so arranged as to take up as little room as possible in the warehouse. (Bacon, 1620/1968, pp. 254-255)

Bacon wanted experimental philosophers to write in plain, simple sentences without recourse to metaphor or other figures of speech. Experimental philosophers, such as Boyle, Hooke and Huygens, endorsed Bacon's proposal about the type of language that should be used to describe new discoveries about Nature obtained by using experimental philosophy (Hooykaas, 1987). However, they were well practiced also in using metaphor to communicate their ideas. Metaphors served as powerful linguistic tools to help readers of experimental reports make connections in their thinking between new observations described by the experimenter and familiar ideas or objects known to the reader. Hooke, for example, used metaphor in reports of his observations, as his description of a flea under the microscope demonstrates:

> But as for the beauty of it, the Microscope manifests it to be all over adorn'd with a curiously polish'd suit of sable [black]. Armour, neatly jointed, and beset with multitudes of sharp pinns, shap'd almost like Porcupine's Quills or bright conical Steel-bodkins; his head is on either side beautify'd with a quick and round black eye. (Hooke, 1665/1961, p. 210)

With the establishment of the Royal Society, however, greater emphasis was placed on the use of simple plain language to describe observations of Nature and experimental procedures. Thomas Sprat in his "History of the Royal Society" criticised the use of ornaments of language including metaphor:

> Who can behold without indignation how many mists and uncertainties these specious Tropes and Figures have brought upon our Knowledg[e]? ...And in few words, I dare say that in all the Studies of men nothing may be sooner obtain'd than this vicious abundance of Phrase, this trick of Metaphors, this volubility of Tongue which makes so great a noise in the world. (Sprat, 1667, p. 112)

With the emergence of scientific journals such as *Philosophical Trans-actions*, the official journal of the Royal Society, Bacon's arguments about the type of discourse that was appropriate for the reporting of experimental philosophy became the required discourse for publication in these journals. Over time, this discourse became accepted as indicative of scientific discourse and gave rise to the acceptance of a cultural myth of language.

> Science Cultural Myth of Language 1:
> The language used to describe the outcomes of an experimental activity should be plain and simple, and free of figures of speech.

If you like, these experimental philosophers were in a cleft stick. They appreciated the power of metaphor to help their readers visualise new knowledge that they were generating. However, at the same time, there emerged the notion that metaphor, and other figures of speech, were not appropriate for reporting experimentation.

Escaping from the Ambiguity of Language: Developments in the reporting of science by 17th century experimentalists also led to calls for a language of science that would systematise new knowledge and separate it from folk knowledge (Halliday & Martin, 1993). Two separate approaches developed. One approach led to the development of specific words to describe scientific concepts. Words such as 'acid', 'apparatus', 'laboratory', 'gravity', 'lens' and 'microscope' were invented in the 17th century (Savory, 1967). For words such as 'acid', their meaning became more clearly defined as experimental philosophers learned more about the properties of the substances they described. Thus, the meaning of these newly coined scientific words evolved over time (Milne, 1996).

The other approach was the development of philosophical languages. John Wilkins, who was a populariser of new scientific ideas and one of the identified founders of the Royal Society in England, proposed the development of a philosophical language which consisted of a "distinct mark to represent every thing" (Vickers, 1987). He proposed the elimination of national languages and the implementation of his philosophical language. According to Wilkins (Vickers, 1987), his language was superior because each object and phenomenon in the universe could be identified by a specific and distinct "mark", which removed the problem of ambiguity that existed between words in the vernacular and the natural object. Wilkins' arguments struck a chord with experimental philosophers who distrusted words. By conceptualising a direct relationship between word, meaning and object Wilkins argued for the need to develop 'correct' definitions. With the invention of new words, experimental philosophers also accepted the need to define terms clearly. An outcome of positing the need for correct definitions is that once these definitions are applied people start to believe that a particular scientific term has a secure meaning. One can understand also the attraction of correct definitions to 17th century realists if it meant there could be a direct relationship between a word and an object.

However, although the language of science, which constitutes a literary discourse, does not undergo the rapid changes of oral discourse, meanings for words used in science are not fixed immutably and do change with the generation of new knowledge and changes in conceptualisation. One of the characteristics of myth is that historic constructs lose the memory that once they were made. Thus, we have the emergence of a second cultural myth related to our memories of scientific language development.

> Science Cultural Myth of Language 2:
> In science there is a one-to-one relationship
> between a natural object, a word for describing it,
> and the meaning of the word.

Consequently, the language of science evolved into a formal discourse based on the imperatives of the myth of realism for simple, unadorned language to describe experimentation, the need for new words to constitute the grammar of science, and the need for rules about how and where these words could be used in the construction of discourse.

Data Speak for Themselves: With their use of the *first person* and *active voice* to convince readers that they were reading an exact record of their experiments, the detailed reports of 17th century experimental philosophers bear a striking similarity to ethnographic reports of contemporary cultural anthropologists. This is evident in Boyle's account of his experimental procedure:

> Nor was I content with this, but for your Lordships further satisfaction and my own I caused a mouse that was very hungry to be shut in [inside the airpump] all night, with a bed of paper for him to rest upon. And to be sure that the receiver was well closed I caused some air to be drawn out of it, perceiving that there was no sensible leak I presently readmitted the air at the stop cock, lest the want of it should harm the little animal[.] (Boyle, 1660/1965, p. 99)

The reports of both Boyle and Hooke indicate their need to convince their readers that they would not allow self-interest to influence their objectivity in the conduct of experimental philosophy. Their use of narrative style in the description of the actual procedure is characteristic of Clandinin and Connelly's *personal experience method* (1994) of reporting educational research which illustrates the firsthand participation of the researcher in the research. Like Boyle and Hooke, Clandinin and Connelly use narrative to present evidence of human experience and to present an account of what happened from the perspective of the participant in order to convince the reader of the legitimacy of the researcher's knowledge claims.

However, a major point of contrast between personal experience method and the experimental philosophy of Boyle and Hooke is that, in the latter case, the data, not the observer, was the controller of the results. Consistent with naive realism, the observer did not 'construct' the data because the data came directly from God's creation, Nature. The observer collected data from Nature which he reported to others through an official language which was thought to be 'transparent': meaning 'shone' directly through the open window of language. It was assumed that the readers' conceptions of matters of fact and reality would be identical to the writer's. Experimental philosophers believed that a direct relationship existed between Nature, observations, matters of fact and the readers' conceptions of the aspects of Nature described in the experimental report.

As the experimental report gradually became accepted as the official method of presenting scientific discoveries, the active voice was removed from accounts of method and results and the pre-eminence of the data over the observer was reinforced. This development, combined with the distancing and non-emotive nature of scientific grammar, influenced the way experimental philosophers wrote about their discoveries, especially after Newton. The implication of this objectivist style is that the data 'spoke for itself' and that if the data were described in a non-emotive way then all readers would develop exactly the same conceptions about newly discovered knowledge of the natural world. The descriptive nature of experimental report writing was accepted but its rhetorical nature was ignored.

Science Cultural Myth of Language 3:
Language is transparent and has no influence
on the interpretation of data generated from
observations of the natural world.

***Structuring the Knowledge*:** At the same time that these cultural myths about grammar and style of language were evolving, a debate was taking place about the appropriate way of 'presenting' new scientific knowledge in experimental philosophy. While Boyle in England was developing a proto-typic form of the 'experimental report', argumentative disputation as a way of presenting new knowledge was very popular in the Italian courts. In Italy, Galileo was the foremost practitioner of these disputations that he used to support his claims to have discovered new knowledge by experimentation. His *Dialogue Concerning Two Chief World Systems,* published in 1632, was characteristic of this approach. The historian Mario Biagioli (1992) attributes this type of scientific discourse to the socio-political structure of the feudal environment in which Galileo worked. He was employed by an absolute ruler who could decide how Galileo's scientific arguments about his discoveries were to be presented to the public and could select the method of presentation ranging from a spectacle to a fictional narrative.

In contrast, the form of experimental report which emerged in England was shaped by a socio-political environment in which there were no absolute rulers, once they had been removed by the Civil War, and where social conventions and politeness were considered to be important aspects of the presentation of any argument, including scientific arguments. Boyle's accounts of his experiments reveal that he adopted a style of report writing that was consistent with Bacon's recommendations in his treatise *Parasceve*. Bacon recommended, firstly, that a legitimate report of an experimental investigation should "diligently and exactly set forth the method" so that the readers could conduct the experiment themselves if they so desired (Bacon, 1968/1620, p. 252). Secondly, the report must contain proof that the results of the experiment were witnessed. In contemporary scientific studies this has been replaced by obtaining multiple sets of results. Bacon added that if the conduct of the experiment was clearly described then "men may be free to judge for themselves whether the information obtained from that experiment be trustworthy or fallacious" (p. 261). Next, Bacon claimed that the experimenter should honestly describe "anything doubtful or questionable" (p. 261) and not try to suppress this information. Finally, Bacon argued that the report should finish with a discussion of the significance of the results and implications for theory of the study. In the account below, we note Boyle's attempt to convince his readers that his experiments were conducted in the manner that he described. He provided enough detail so that they could either repeat the experiments or be able to imagine the reality of the experiment and be convinced of the experimental results. Boyle writes that he hopes that people:

> [M]ight, without mistake, and with as little trouble as possible be able to repeat such unusual experiments; and that after I consented to let my observations be made public, the most ordinary reason for my prolixity was that, foreseeing that such a trouble that I met with making those trials carefully, and the great expense of time that they necessarily require... will probably keep most men from trying again these experiments I thought I might do the generality of my readers no unacceptable piece of service by so punctually [carefully] relating what I carefully observed, that they may look upon these narratives as standing records of our new pneumatics. (Boyle, 1660/1965, p. 2)

However, it was with the publication of the journal of the Royal Society, *Philosophical Transactions,* that this structure for reports of experimental activity was adopted more widely. This prestigious journal defined both the language that was appropriate for experimental philosophers to use in their reports and the structure that the reports should have if they were to be published in the journal. With both these requirements, the publishers were heavily influenced by the writings of Bacon. Experimental reports were constructed, therefore, within a particular social milieu. Legitimation of knowledge was a public endeavour and accounts had to be convincing in their accuracy, believability and veracity. The unchanging nature of the experimental report has resulted in its structure influencing conceptions of how the experiment, described in the report, had been conducted. When it became accepted as a 'matter of fact' that scientific experiments are performed in the same linear fashion that is inherent in the structure of the experimental report, a new cultural myth was born. Experimental activity came to be defined by the experimental report, and vice versa (Dear, 1991).

Science Cultural Myth of the Experimental Report: The process of scientific endeavour is represented by the specific structure of the experimental report.

In summary, the experimental report developed at the same time as the emergence of powerful learned societies, such as the Royal Society in England, and their promotion of it ensured that power was retained by this genre in the reporting of scientific discoveries. From within the myth of realism, the experimental report constitutes an objective account of true knowledge of the natural world. The carefully contrived impersonality of the official language of the report renders the language transparent. And, when a skilful author (whose rhetorical skills remain unacknowledged) meets the literary standards of the genre, it is as though the experimental report is a window into Nature.

From a critical constructivist perspective, however, cultural myths that evolved from the arguments of Bacon and his like-minded believers captured the experimental report and wrought it as a cultural artifact so that it became an object. This genre became a discourse of power that permitted one method for legitimating scientific knowledge. Over time, the framing presence of the myth was lost and this discourse became the inevitable, almost sacred, discourse of science.

Rather than a window into Nature, the language of the experimental report is a powerful rhetorical tool designed to present a specific argument (Cantor, 1989). We argue that disempowerment results from failure to appreciate the historic role of the myth of realism in framing the rhetorical form and style of the literary genre of the experimental report. The danger inherent in blind acceptance of the imperatives of the experimental report is the likelihood of believing that a report 'presents' results and conclusions rather than accepting that it constitutes a representation of its author's knowledge. In rejecting the metaphor of language as a window, we argue that language must be examined as an important participant, along with observation, Nature and experimentation and facts, in the pursuit of the construction of science.

Pedagogical Myths of Language: Realist beliefs in the inappropriateness of metaphor and other expressive forms of language, a primary emphasis on definitions, and removal of first person accounts and active voice have removed the human face from contemporary school science and replaced it with a collection of objects.

The need in science to define carefully often is perceived in science teaching as a fundamental step towards student understanding. Definitions abound in many science texts. Where there is a primary pedagogical emphasis on learning scientific facts there also seems to be a corresponding emphasis on students re-presenting correct scientific definitions cited by learned sources such as the teacher or the textbook. The application of a single meaning for each word has tended to be emphasized in science teaching to the extent that the meanings of words are presented as unchanging. Consequently, specific words that have been developed to describe scientific concepts (e.g., acid, molecule, field) often are presented without reference to their historical development.

If teachers believe that meanings are fixed and timeless then words become identified with a specific object or idea. If there is a perceived one-to-one relationship between a word and an object then the need for the teacher or the textbook to adopt the role of informer is reinforced because only the teacher and the text have access to the correct meanings. This can lead to an undue emphasis in the classroom on defining correctly ideas and objects, and can reinforce the notion of the simplistic nature of the language of science. In practical activities, the notion of a fixed relationship between object and word can lead teachers and students to assume that descriptions of observations generate identical pictures of reality in the minds of all participants.

> Pedagogical Myth 4:
> Throughout time, scientific definitions are fixed. The teacher/textbook is the source of scientific definitions. The relationship between object and word is fixed.

In a similar way, the language of school science, especially of practical reports, is governed by and, in turn, promotes implicitly the myth of the transparency of language. From a realist perspective, language is analogous to a pane of glass through which descriptions of reality pass from one reader to the next without distortion.

> Pedagogical Myth 5:
> In the practical activity, data speaks for its self; and because language is transparent, language serves as a vehicle for accurately describing reality.

As an icon of the scientific process, the experimental report can remain taken-for-granted in the classroom. Often it is used to present a highly structured account of student practical activities as a corollary to the way experimental reports are used to report new discoveries in science. As we have argued, this structure has its antecedents in the writings of Francis Bacon and in the demands of the Royal Society. Usually, students are required to write their reports totally in the passive voice, thereby removing completely the student as the active agent and emphasizing the object, that is, the practical activity. For students, the message implicit within this practice is that they are less important than the activity.

Pedagogical Myth 6:
The experimental report is a descriptive account of a practical activity. It is a true record of experimental activity.

In science, the research paper, or experimental report, has become an accepted way of arguing for the legitimation of purported discoveries. However, there are two important features of contemporary experimental reports that science educators need to understand better if they are to avoid perpetuating unwittingly the above pedagogical myths. First, doing science and reporting science are distinctive activities. In their studies of science laboratories, Latour and Woolgar (1979) have argued that the practice of science is far removed from the reporting of science. Second, there is scope for writing experimental reports in multiple voices. According to Riley (1991), the contemporary experimental report contains both passive and active voice. For example, active voice can appear in the Introduction section, where the author is chronicling and critiquing previous research, and in the Discussion section, where the author is interpreting the data and advocating new theory.

So, contemporary scientific experimental reports can have a rhetorical character similar to that of 17th century experimental reports. It seems ironic to us that, although Bacon's inductivism was recognized eventually as not a true reflection of the methods of science (Hooykaas, 1987), his proposal for experimental reports continues to dominate the discourse of school science. Although there has been a recent move in the social sciences toward the use of narrative, emotive and active rhetoric, in school science and in the mass media, the discourse of science continues to be represented overwhelmingly as non-emotive, distancing and passive.

From our critical constructivist perspective, an exclusive emphasis in the science classroom on the use of the passive voice can disempower students because it removes from them a sense of agency as active constructors or interpreters of the data of their practical activities. If students are encouraged to be more active and more rhetorical in their reporting then they are likely to experience an enhanced sense of involvement and gain greater enjoyment from their practical activities. Why not provide writing opportunities for students to celebrate the excitement and intrigue of their personal experiences as they peer down microscopes and observe for the first time the alien inhabitants of micro-worlds? Why not encourage them to pretend to be early natural historians so struck by the beauty of Nature that florid prose and poetic expression are appropriate rhetorical devices to evoke rich images in the minds of their readers? We believe that it is important for school science to present the language of science as a vibrant and evolving aspect of scientific endeavour, rather than as a series of impersonal *a*historical definitions.

CONCLUSION

If constructivist reform of science education is to be successful, science educators need to deconstruct some formidable barriers whose strength lies in their current invisibility. By adopting a critical constructivist perspective, we can recognise the historio-cultural nature of these barriers, or cultural myths, and their role in perpetuating many of the taken-for-granted discursive practices of contemporary school science. In this chapter, we have focused largely on pedagogical myths that govern the practical activity of modern school science. We have traced the roots of these science-teaching myths to the development of experimental science in the seventeenth century.

The ontology, epistemology and methodology of experimental science evolved in a robust intellectual environment of competing possibilities. The eventual predominance of realism owes much to the politics of legitimation associated with seventeenth century learned societies, their royal patronage, and a widespread belief in natural theology. A decisive commitment to realism gave rise to other powerful commitments, including (1) belief in the primacy of observation and the privileged status of scientific facts, and (2) a value-neutral literary genre of scientific reporting that eschewed metaphor and ambiguity while masquerading as a metonym for scientific activity. In time, these powerful beliefs became enshrined as pervasive cultural myths whose apparent naturalness and inevitability was reinforced by the invalidation of alternative discursive practices and a loss of memory in their own historical contingency. In other words, humanity forgot that it had constructed an edifice and, instead, took it as ordained by fate, God or the natural superiority of the Western mind.

Science education of the twentieth century will be remembered not only for its role in perpetuating the cultural myths of seventeenth century experimental science, but also for the revolutionary role of constructivism in contesting the epistemological stranglehold of these 300-year-old myths on the practices of school science. Although constructivism has been partially successful in contesting some pedagogical myths (e.g., transmissionism) which underpin the apparent naturalness of unilateral teacher control of the learning environment, it has lacked the conceptual power to contest their ontological and methodological dimensions. Now, critical constructivism offers a much-needed moral basis for doing so. By combining a critique of ideology with an appreciation of historical processes, critical constructivism provides a framework for enabling science educators to identify and transform these historio-cultural myths, particularly those which ignore the learner in school science.

From a critical constructivist perspective, we envisage the emergence of school science classroom environments in which students are empowered through critical reflective thinking to understand the historical and cultural contingency and interconnectedness of the discursive practices of both science and school science. While being enculturated into scientific ways of exploring and making sense of the natural world, students should, at the same time, be made critically aware of the prevailing framework of enculturation. School science should become a rich educative forum in which students learn both to apply (pluralistic) scientific standards of judgement and to be critically aware of the standards of judgement underpinning all knowledge claims, including those framed by both realist and constructivist belief systems.

References

Arbib, M. A., & Hesse, M. B. (1986). *The construction of reality*. Cambridge: Cambridge University Press.

Bacon, F. (1968). Paraceve/Novum Organum. In J. Spedding, R. L. Ellis, and D. D. Heath (eds.), *The works of Francis Bacon*. New York: Garrett Press, (Original publication 1620, facsimile reprint of 1870 publication).

Bakhtin, M. M. (1981). *The dialogical imagination*. Austin: University of Texas Press.

Barthes, R. (1972). *Mythologies*, A. Lavers (trans.). New York: Hill & Wang.

Biagioli, M. (1992). 'Scientific revolution, social bricolage, and etiquette', in R. Porter and M. Teich (eds.), *The scientific revolution in national context*. Cambridge: Cambridge University Press, 11-54.

Bowers, C. A. (1987). *The promise of theory: Education and the politics of cultural change*. New York: Teachers College Press.

Boyle, R. (1965). 'New experiments physico-mechanical touching the spring of the air; Made for the most part, in a new pneumatical engine', in T. Birch (ed.), Robert Boyle: *The works Vol. 1-6*, Georg Olms Verlagsbuchhandlung, Hildesheim, 1-117. (Original publication 1660, facsimile reprint of 1744 publication).

Britzman, D. P. (1991). *Practice makes practice: A critical study of learning to teach*. Albany, NY: State University of New York Press.

Bruner, J. (1986). *Actual minds, possible worlds*. Cambridge, MA: Harvard University Press.

Campbell, J. (1972). *The hero with a thousand faces*. Princeton, NJ: Princeton University Press.

Cantor, G. (1989). 'The rhetoric of experiment', in D. Gooding, T. J. Pinch, & S. Schaffer (eds.), *The uses of experiment: Studies of experimen-tation in the natural sciences*, Cambridge: Cambridge University Press, 159-180.

Clandinin, D. J. & Connelly, F. M. (1994). 'Personal experience methods', in N.K. Denzin and Y.S. Lincoln (eds.), *Handbook of qualitative research*. Thousand Oaks, CA: Sage Publications, 413-427.

Crosland, M. (ed.) (1976). *The emergence of science in Western Europe*. New York: Science History Publications, 1-13.

Dear, P. (1991). 'Narratives, anecdotes and experiments: Turning experience into science in the seventeenth century', in P. Dear (ed.), *The literary structure of scientific argument*. Philadelphia: University of Pennsylvania Press, 135-163.

Descartes, R. (1970). 'Third set of objections and replies containing the controversy between Hobbes and Descartes', in E. Anscombe and P. T. Geach (eds.), *Descartes philosophical writings*, Nelson's University Paperbacks, Sunbury-on-Thames, Middlesex, 127-150 (Based on the original published 1642).

Douglas, M. (1967). 'The meaning of myth', in E. Leach (ed.), *The structural study of myth and totemism*. London: Tavistock Publications, 49-69.

Durkheim, E. (1976). *The elementary forms of religious life*. London: Allan Unwin.

Erickson, F. (1986). 'Qualitative methods in research on teaching', in M.C. Wittrock (ed.), *Handbook of research on teaching* (3rd ed.). New York: Macmillan, 119-159.

Findlen, P. (1993). 'Controlling the experiment: Rhetoric, court patronage and the experimental method of Francisco Redi', *History of Science, 31*, 35-64.

Freud, S. (1965). *The interpretation of dreams*. New York:Avon.

Gaarder, J. (1995). *Sophie's world*, P. Moller (trans.). London: Phoenix House.

Geertz, C. (1973). *The interpretation of cultures*, New York: Basic Books.

Guba, E. G. & Lincoln, Y. S. (1989). *Fourth generation evaluation*. Newbury Park, CA: Sage Publications.

Habermas, J. (1972). *Knowledge and human interests*. London: Heinemann Educational Books.

Habermas, J. (1978). *Legitimation crisis*, T. McCarthy (trans.). Boston: Beacon Press.

Habermas, J. (1984). *A theory of communicative action: Vol 1. Reason and the rationalisation of society*, T. McCarthy (trans.). Boston: Beacon Press.

Halliday, M. A. K. & Martin, J. R. (1993). *Writing science: Literacy and discursive power*. London: The Falmer Press.

Hardy, M. D. & Taylor, P. C. (1997). 'Von Glasersfeld's radical constructivism: A critical review', *Science & Education, 6*, 135-150.

Helu, I. F. (1994). 'Mythical and scientific thinking: A comparison', in J. Edwards (ed.), *International interdisciplinary perspectives*. Victoria: Hawker Brawhlow Education, 66-72.
Hobbes, T. (1970). 'Third set of objections and replies containing the controversy between Hobbes and Descartes', in E. Anscombe and P. T. Geach (eds.), *Descartes philosophical writings*. Sunbury-on-Thames, Middlesex: Nelson's University Paperbacks, 127-150. (Based on the original published 1642).
Hooke, R. (1961). *Micrographia*. New York: Dover Publications. (Facsimilie copy of original published 1665).
Hooykaas, R. (1987). 'The rise of modern science: When and why?', *British Journal for the History of Science*, 20, 453-473.
Jung, C. G. (1968) *The archetypes and the collective unconscious* (2nd. ed.). London: Routledge and Kagan Paul.
Kuhn, T. (1970). *The structure of scientific revolutions* (2nd. ed.). Chicago: University of Chicago Press.
Lakoff, G. & Johnson, M. (1980). *Metaphors we live by*. Chicago: University of Chicago Press.
Latour, B. & Woolgar, S. (1979). *Laboratory life: The social construction of scientific facts*. Beverley Hills: Sage Publications.
Levi-Strauss, C. (1970). *The raw and the cooked*, J. Weightman and D. Weightman (trans.), Jonathan Cape, London.
Malinowski, B. (1954). *Magic, science and religion and other essays*. Garden City, NY: Doubleday Anchor.
Malinowski, B. (1971). *Myth in primative society*. Westport, CN: Negro Universities Press.
Matthews, M. R. (1993). 'Constructivism and science education: Some epistemological problems', *Journal of Science Education and Technology*, 2, 359-370.
McCarthy, T. (1978). *The critical theory of Jurgen Habermas*. London: Polity Press.
Milne, C. (1996). *The representation of 'acid' in school chemistry: From concept to fact*. Paper presented at the 14th International Conference of Chemical Education, Brisbane, Australia.
Osborne, R. & Wittrock, M. C. (1983). 'Learning science: A generative process', *Science Education, 67*, 489-508.
Polanyi, M. & Prosch, H. (1975). *Meaning*, University of Chicago Press, Chicago.
Pusey, M. (1987). *Jurgen Habermas*, London: Tavistock.
Reddy, M. J. (1979). 'The conduit metaphor – A case of frame conflict in our language about language', in A. Ortony (ed.), *Metaphor and thought*. Cambridge: Cambridge University Press, 284-324.
Riley, K. (1991). 'Passive voice and rhetorical role in scientific writing', *Journal of Technical Writing and Communication, 21*, 239-257.
Savory, T. (1967). *The language of science*. London: Andre Deutsch.
Schon, D. A. (1983). *The reflective practitioner: How professionals think in action*. New York: Basic Books.
Shapin, S. & Schaffer, S. (1985). *Leviathan and the air pump: Hobbes, Boyle, and the experimental life*. Princeton, NJ: Princeton University Press.
Shapin, S. (1984). 'Pump and Circumstance: Robert Boyle's literary Technology', *Social Studies of Science, 14*, 481-520.
Slaughter, M. M. (1982). *Universal languages and scientific taxonomy in the seventeenth century*. Cambridge: Cambridge University Press.
Slotkin, R. (1973). *Regeneration through violence: The mythology of the American frontier, 1600-1860*. Middletown, CN: Wesleyan University Press.
Smolicz, J. J. & Nunan, E. E. (1975). The philosophical and sociological foundations of science education: The demythologising of school science, *Studies in Science Education, 2*, 101-143.
Sprat, T. (1959). *History of the Royal Society*. St. Louis: Washington University Studies. (Facsimile of original publication 1667).
Sutton, C. R. (1992). *Words, science and learning*. Buckingham: Open University Press.
Sutton, C. (1994). '"Nullius in Verba" and "Nihil in Verbis": Public understanding of the role of language in science', *British Journal for the History of Science, 27*, 55-64.
Taylor, P. C. (1996). 'Mythmaking and mythbreaking in the mathematics classroom', *Educational Studies in Mathematics, 31*, 151-173.
Taylor, P. C. (in press). 'Constructivism: Value added', in B. Fraser & K. Tobin (Eds.), *The international handbook of science education*. Dordrecht, The Netherlands: Kluwer Academic Press.

Taylor, P. C. & Williams, M. C. (1993). 'Critical Constructivism: Towards a Balanced Rationality in the High School Mathematics Classroom', Paper presented at the annual meeting of the American Educational Research Association, Atlanta, GA.

Tobin, K., Davis, N., Shaw, K. & Jakubowski, E. (1991). 'Enhancing science and mathematics teaching', *Journal of Science Teacher Education, 2*, 85-89.

Toulmin, S. (1990). *Cosmopolis: The hidden agenda of modernity*. New York: The Free Press.

Vance, K. & Miller, K. (1995). 'Setting up as a constructivist teacher: Examples from a middle secondary ecology unit', in B. Hand & V. Prain (eds.), *Teaching and learning in science: The constructivist classroom*. Sydney, Australia: Harcourt Brace, 85-105.

Vandenberg, D. (1990). *Education as a human right: A theory of curriculum and pedagogy*. Columbia University, NY: Teachers College Press.

Vickers, B. (1985). 'The Royal Society and the English prose style: A reassessment', in B. Vickers & N. Struever (eds.), *Rhetoric and the pursuit of tuth: Language change in the seventeenth and eighteenth centuries*. Pasadena, CA: Castle Press, 1-76.

Vickers, B. (1987). *English science, Bacon to Newton*. Cambridge: Cambridge University Press.

von Glasersfeld, E. (1990). 'Environment and communication', in L.P. Steffe & T. Wood (Eds.), *Transforming children's mathematics education*. Hillsdale, New Jersey: Lawrence Erlbaum Associates.

von Glasersfeld, E. (1991). 'An exposition of constructivism: Why some like it radical', *Journal for Research in Mathematics Education Monograph 4*, 19-29

Gili S. Drori

Chapter 3

A Critical Appraisal of Science Education for Economic Development

No country has been successful economically with less than 50% literacy.
Katherine Marshall[19]

... chances are [children from all around the world] will rely on the mathematics and science they learned in this decade to succeed in the complex business and technological environment of 2012.
Learning Science (1992, p. 4)

Expert advice and lay-notions regard science as an essential social institution: science is a necessary element of modern education, it is a building block for personal and social development, and its products advance human society and offer prosperity. Science is also considered as an integral part of modernity, of advanced nations and of enlightened and civilized societies. We judge nation-states by their scientific performance or by their utilization of technology. We compare societies by their level of scientific literacy or by the number of great inventors from their ranks. We observe that the more advanced societies are those with most advanced science institutions, and, thus, we conclude that science and national progress are causally linked. Scientific progress rests on scientific knowledge and scientific skills; hence, science education is essential to future scientific advances and to, in turn, economic prosperity. Consequently, in our everyday life we assign science education the social role of bringing progress to society, especially to the most needy of societies. This social role of science education is taken-for-granted: we assume it as such, we act upon it, and rarely do we question our opinion of science education. I suggest that we pose to reassess our standing towards science education and towards the social role that we assign to it. In this chapter I offer some evaluations of our everyday notion of science education and of the translation of this everyday belief into social policy.

This chapter is composed of three main parts and a concluding section. Part I offers a description of the modern concept of science education: Why do we praise the venture of science? What are the expected benefits from science education?

---

[19] Katherine Marshall is an Africa specialist with The World Bank and is quoted in the *Wall Street Journal* (27 January 1994).

W. W. Cobern (ed.), Socio-Cultural Perspectives on Science Education, 49–74.

What is the model for the social role of science education? How is such a model translated into the realm of action, in national and international policies? In summary, I describe the conceptual model of "science education for development", which focuses on the national economic benefits from scientific activity and investment. This conceptual model is embedded in policy discussions, whether development-, education-, or Third World oriented policies. In addition, the diffusion of this conceptual model is greatly encouraged by the UN and its dependent agencies, such as UNESCO. Part II of this chapter provides a critical assessment of this conceptual model: What is the empirical evidence to refute the model? What theoretical explanations are provided for such empirical findings? Hence, I review ample empirical research that supports, or refutes, the claims made in the “science education for development” model. In this sense, this Part provides a realist critique, i.e., a critical evaluation of the conceptual model, while accepting its general premise – science education leads to national development. Part III offers a critique of the "science education for development" model from "outside" of the realist framework. In it, I question the discursive framework which links science education and progress, examine its related policies as discursive narratives, and provide an explanation for the historical construction of this model. I argue that science – the national commitment to science education, policy, and investment – is a ritual of modern nation-statehood and is dependent on the world polity, its organizational structure and the global discourse. Part IV concludes my institutionalist assessment of the field of comparative science education by offering additional directions for further research.

Part I: Science Education and National Development - Establishing A Causal link

1.1 The Model of "Science for development": Concepts and Policy Guidelines

Science, including science education, are highly praised as a requirement for any modern, civilized, economically vibrant society. Like education, science sets the basis for several paths towards national development. While some researchers and policy-makers take a normative approach to national development[19], most discussions rest on a structural, instrumentalist perspective. According to the structural-instrumentalist approach, national economic growth depends on the scientific and technical capabilities of the labor force; such capabilities rely on the level of advanced scientific and technical training; and, finally, such advanced

[19] The normative approach to national development argues that progress relies not on structural factors, but rather on the gradual change in the nature of the people who together compose nation-states. Accordingly, social change is caused by the change in the normative orientation of the members of society. It is the values, motivations and psychological forces that bring societal changes about. While this approach draws extensively from Weber's classical study "The Protestant Ethic and the Spirit of Capitalism" (1904), it furthered such discussions to issues of the characteristics of "modern man" (e.g., Inkeles & Smith 1974), and the components of the "mental virus" and the processes by which it "transmits" modernity (e.g., McClelland 1961, 1969.) Regarding the role of science in national development, the effect of science – positive or negative – is mediated by the normative position that is carried by science. Science is the carrier of a rational, cause-and-effect worldview, which is based on Western logic – what is called "the scientific mind." Thus, it is such normative changes in the indigenous people, who are now science educated, that lead to national development. As Kelly (1990, p. 53) phrases, "the idea is not merely to provide education in science, but education through science."

training rests on the foundations of science education in primary- and secondary schools. I summarize this conceptualization of the social role of science into the model of "science for development." Figure 1 is a graphic display of the tenets of the "science for development" conceptual model.

The *science-for-development* model establishes a causal, serial link among science schooling, advanced or 'applied' science, and the economic conditions of a nation. It regards science education as a means, or a mechanism, for achieving the objective of national development. This model elaborates on the notion that science education and economic progress are mediated by technology,[20] by specifying that it is the technical capabilities of local personnel and the technological products created by such skilled laborers that stand for what was previously generalized as "technology." Furthermore, implied in this conceptual model is that specific scientific disciplines are expected to contribute to the development of their relevant economic sector. For example, medical sciences affect health conditions, physics lead to the development of atomic energy sources, engineering and geo-physics contribute to solving problems of water resources, etc.[21] Finally, a national emphasis on specific scientific disciplines may also provide a competitive edge to the local economy. For example, African countries can emphasize zoology and botany, thus relying on a rare and unique local resource and providing an opportunity for increased tourism and related revenue.[22]

This model has four major assumptions. First, this model regards science as a *national* project, i.e., a scheme which is aimed at providing benefits for the nation and which relies on national financial support and societal legitimacy. Second, it envisions a particular plan, or systemic program, for the achievement of national development. Unlike national policies for support of, for example, the arts, the "science for development" plan is very explicit in its vision of the path which leads to economic development: science education leads to scientifically- and technically skilled labor force, and such skilled personnel enables industrialization and economic progress. Third, it considers science to be a "real" social institution, rather than a socially constructed phenomenon.[23] And, last, the model allows for the discursive regime of "development" to dominate any discussion of science, science education, and their social role. Hence, for example, science education is commonly mentioned in policy guidelines for social and economic development and science matters are frequently handled by governmental agencies whose primary concern is national development. In other words, "development" is the master narrative for any

---

[20] For explanations of, examples for, and critical assessment of the hierarchical model among science, technology and the economy, see Drori 1993.

[21] For an example of a discipline-specific approach to national development, see Nayar's (1976) study of India.

[22] This discipline-specific policy for the tertiary sciences, while aiming at carving a viable niche for distressed economies, was not carefully implemented. During the 1970s, for example, African nations, hoped to enhance their competitive edge, yet did not focus on scientific disciplines that directly utilize their national resources. Hence, they followed policy recommendations to adopt the discipline-specific strategy by investing predominantly in social sciences and agriculture. For example, in 1977 Malawi invested 40% of its university public funds in agriculture and 60% in social sciences, and Niger invested 40% in 1976 in social sciences and 9% in agriculture.

[23] The issue of realist and constructivist perspectives on science and development is revisited in section 2.4.

science education issue.[24] This trend parallels an increase in development related arguments in education policy and educational aims during 1950-1970 period (Fiala & Gordon-Lanford, 1987). Moreover, the vision of national development is reduced to *economic* development, i.e., the type of national development that is easiest to quantify and easiest to monitor.[25] In summary, the "science for development" conceptual model promotes a development-oriented, economic-centered, utilitarian vision of science (See Figure 1).

The *science-for-development* model rests on the theoretical foundations of structuralist-functionalist modernization theory, which is essentially a liberal, socio-evolutionist perspective. Whether viewing social development as uni-directional progression (for example, Toennies' studies of community and society or Durkheim's studies of organic and mechanic social solidarity) or whether viewing it as a continuous process (for example, Parsons' definition of pattern variables), all advocates of modernization theory envision an "ideal type" of society. When applied to the study of nations, modernization theory conceptualizes national development as divided into *stages*. Hence, the condition of the developed countries is regarded as a higher state, or a higher level, of achievement for nation-states, and the condition of Third World countries as merely a lower, more backward stage. It further supposes that situations and events are transferable from one nation-state to another. Thus, modernization theory suggests that less developed countries (LDCs) follow in the footsteps of developed countries (DCs), in order for LDCs to achieve the desired economic prosperity and social liberties. In this scheme, science and education are envisioned as the mechanisms that proved effective in the progress of DCs, and, thus, are expected to ripe similar fruits in LDCs.[26]

1.2 Science Education for Economic Development

In accordance with modernization and human capital theories, science education was institutionalized worldwide, and especially adopted in LDCs. Much like the notion of mass schooling as a means for creating a brighter future for LDCs through the modern socialization of their future citizens, so did LDCs put their faith in science education. While mass schooling carries the idea of the individual as a citizen of a modern nation-state (Meyer 1987, Ramirez & Rubinson, 1987, Meyer, Ramirez & Soysal, 1992), science education further specifies the notion of the modern citizen as a rational, technically-able, environmentally-conscientious person.[27] Our notion of

---

[24] For an analysis of the international discourse of development and an example of how this reductionist discourse forces countries such as Lesotho into a uni-dimensional category of LDCs, see Ferguson 1990.

[25] Even the normative approach to development (which, as mentioned earlier, regards science as promoting values and motivations of modernity) or the "new model of development" (which identifies the objectives of national development as both economic and cultural; Mayor 1982) are focused, at the last instance, on economic results. In other words, these approaches identify a factor – values of modernity or a combination of cultural and economic elements – as mediating between science and national economic development.

[26] Modernization theory inspired policy guidelines regarding science (e.g., Moravcsik 1966, 1971) and literacy and mass education (e.g., Golden 1955, Bowman & Anderson 1965.) See Moravcsik (1987) for a more explicit theoretical positioning.

[27] The theme of rationality is historically a component of science education. The themes of technical ability and environmental conscience were added later to the core theme of rationality. See further in McEneaney 1995.

science education narrows the general objectives of mass schooling and directly links between education and national progress. It is assumed, hence, that mathematics and science curriculum[28] have greater economic consequences than other instruction fields of mass education, such as classical education, humanities or languages.

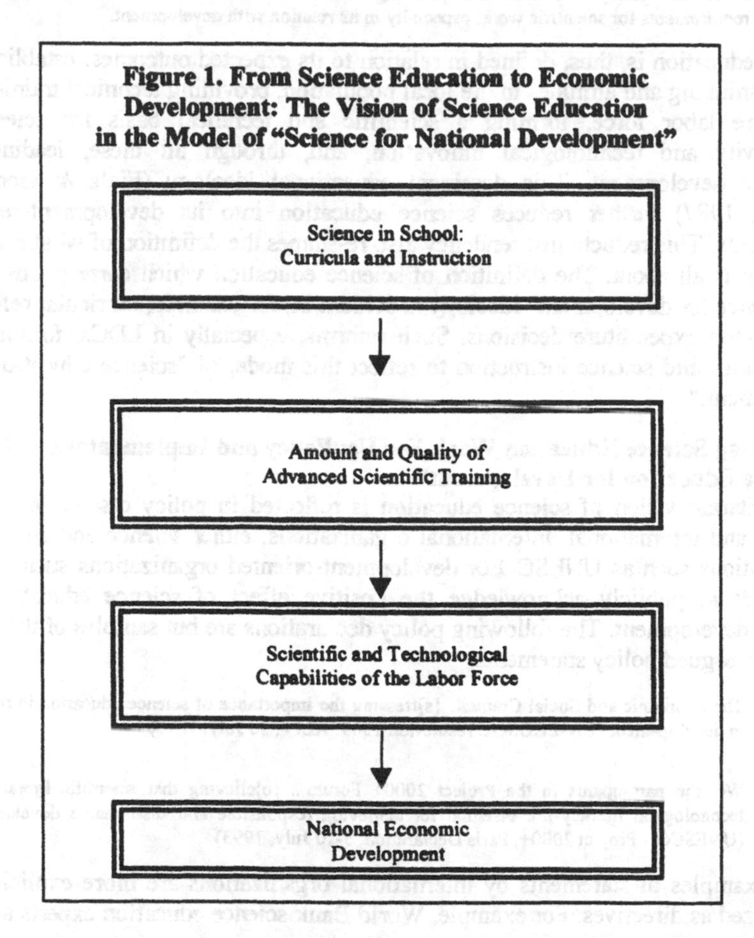

Figure 1. From Science Education to Economic Development: The Vision of Science Education in the Model of "Science for National Development"

In this scheme of "science education for development", the role of science education is two-fold: (a) to shape positive attitudes towards modernization, and (b) to train candidates in science and technology and prepare them for higher education and for sophisticated production roles. In other words, in addition to the general objective of mass education as socializing modern individuals into participatory civil and political life, science education takes the additional role of professional training and providing the local economy with skilled labor. Appropriately, any policy discussion regarding the shortage of adequately skilled personnel or of local

---

[28] It is important to note the curricular distinction between general science issues and science studies for scientists. In primary- and secondary education, however, there is no such differentiation.

innovation, concludes with reasserting the importance of scientific and technological training as part of children's schooling. As Rene Maneu, the director general of UNESCO, states (1971, p. 5):

> **These considerations [the problems of inadequately trained manpower; GD] lead to the linking of science policy with education planning. The latter has to take account of the staffing requirements for scientific work, especially in its relation with development.**

Science education is, thus, defined in relation to its expected outcomes: establishing modern thinking and attitudes in the local population, providing technical training to the future labor force, forming a scientific and technical basis for scientific productivity and technological innovation, and, through all these, leading to economic development. This dominant educational ideology (Fiala & Gordon-Lanford, 1987) further reduces science education into its development-related components. This reductionist tendency also re-shapes the definition of what science education is all about. The definition of science education which corresponds with the "science for development" ideology is evident in, for example, curricular reforms or education expenditure decisions. Such reforms, especially in LDCs, formulated mathematics and science instruction to reflect this model of "science education for development."

1.3 Making Science Education Work For Us: Policy and Implementation of "Science Education for Development"

The utilitarian vision of science education is reflected in policy discussions, both national and international. International organizations, either science and education organizations such as UNESCO or development-oriented organizations such as the World Bank, publicly acknowledge the positive effect of science education on national development. The following policy declarations are but samples of the most explicitly argued policy statements:

> **The Economic and Social Council...[s]tressing the importance of science education in relation to develop-ment (UN-ECOSOC resolution 1309: XLIV, 30 July, 1970)**

> **We, the participants in the Project 2000+ Forum... [b]elieving that scientific literacy and technological literacy are essential for achieving responsible and sus-tainable development. (UNESCO - Project 2000+, Paris Declaration, 5-10 July, 1993)**

Other examples of statements by international organizations are more explicit and are stylized as directives. For example, World Bank science education experts advise that:

> **Governments that are interested in laying the groundwork for a more technically oriented economy... should place heavy emphasis on general mathematics and science... These subjects are relatively inexpensive to teach and are likely to promote economic growth more efficiently than can in-school vocational education. (The World Bank 1988, p. 62)**

Similar statements are echoed in declarations of national policy organizations. For example:

> **The Government sees higher education and the institutes of higher learning as the major means for scientific, technological, and cultural progress, as an important support for the economic and social prosperity of the state. (Basic Principles for Governmental Policy, Government of**

> the State of Israel headed by Mr. Yitzhak Rabin, 1992, section 9.11)
>
> The Council believes that it is of national importance that all Canadians receive a quality education in science and technology. For Canada to cope with social changes rooted in highly specialized technologies, its citizens need the best *general* education possible - an education comprising not only the traditional basics of language and mathematics, but also the *new* basics of our contemporary culture: science and technology. (Science Council of Canada, 1984, p. 9)

Following the 1950s and 1960s worldwide promotion of mass education and of science as means for national development, science education became the next item on the global agenda of national development. The influx of science education policies was inspired by empirical findings as for the positive effects of mass education[29] and of science[30] on national economic development. A few studies provide quantitative support for these policies for science education. Cross-national studies show that tertiary science and engineering enrollments (Ramirez & Lee, 1990) and the share of primary- and secondary school curriculum devoted to mathematics and science (Benavot, 1992) have a significant, positive effect on economic conditions.[31] However, the number of studies devoted to the examination of the particular causal connection between science education and national development is very small and the conclusions drawn from these studies as for the causal link between science education and economic growth are inconclusive. Thus, most of the support arguments for the "science education for development" model comes from case-specific analyses and from general studies.

Analyses of Western, industrialized countries provide examples for LDCs as

---

[29] Overall, mass education has a significant, positive effect on economic growth (this is in spite of the fact that the expansion of mass schooling is independent of differences in intra-national characteristics; Meyer et al. 1977, Ramirez & Boli 1987.) For example, when considering school enrollments Meyer, Hannan, Rubinson and Thomas (1977) find in their cross-national study that primary- and secondary education has a significant positive effect on GNP between 1950 and 1970. Several other studies (Coclough 1982, Heyneman & Siev-White 1985, Lockheed & Vespoor 1990) support the finding that primary school enrollments positively affect economic progress. Ramirez and Lee (1993) observe cross-national positive educational effects (all levels - primary, secondary and tertiary education) on real GDP. In addition, several case studies support these findings. Walters and Rubinson (1983) show that secondary education (as measured by both enrollments and high school degrees) has a positive, twenty-year lag effect on American economic output between 1933-1969. Hage, Garnier and Fuller (1988), while studying France, find that secondary education has positive effects on four-year and ten-year lagged economic outputs between 1950 and 1975. In addition to the general effects, "[qualitative features of national school systems, such as provision of textbooks, per-pupil expenditure, and the extent of teacher training also have modest economic effects, especially in the developing world" (Benavot 1992, p. 150)

[30] Numerous cross-national studies show a significant positive effect of, for example, scientific manpower on economic progress. The number of scientists and engineers (Myers & Harbison 1964, Blute 1972), the number of scientific authors in technical journals (Price 1964 in Inhaber 1977, p. 517), and the number of "publishing scientists" per capita (Inhaber 1977) were found to positively and significantly correlate with, or affect, either economic level or economic growth. In addition, case studies showed that industrial expenditure of R&D in the U.S. positively affected economic conditions (Mansfield 1972, Griliches 1987), and that in sub-Saharan Africa the number of scientific publications positively and significantly correlates with GDP (Zymelman 1990).

[31] These studies do not directly support the "science education for development" model. For example, Benavot (1992), while offering the empirical findings, argues that using this evidence in support of functionalist arguments is a narrow interpretation. He thus sheds a new theoretical light on its meaning, as discussed in Part III.

for the desired connection between science education, industrialization and economic growth. Science education is noted as one of the common elements among the industrialized nations. Moreover, the future of developed countries – their economy, security, global status and "attractiveness to human society" – is believed to depend on successful science education (Rutherford. 1985, p. 207.) It is a widely shared idea that the progress of Western nations – their industrialization and capitalist expansion – is mainly due to technological breakthroughs and to the cumulative scientific knowledge on which these innovations were based. Hence, it seems to establish that science education was, and still is, a major factor in the economic success of the industrialized, prosperous world.

Case studies of the newly industrialized countries (NICs) provide most of the lessons as for the causal link between science education and economic progress. In most of these countries "access to science education is near universal at primary level and the quality of teaching and material resources is becoming comparable to that found in industrialized countries" (Lewis 1993, p. 2). Moreover, science education enjoys high levels of legitimacy and receives high status, relative to other topics (Lewis 1993, p. 3). These growing share and high quality of science education are noted as the causes of the remarkable technology-based economic success of such countries as Taiwan, South Korea, Japan, Malaysia and Singapore (see Altbach, 1989). Even India, in spite of its relative low standards of science and technology education and the lack of strong governmental directives, produced a scientific and technological proletariat, established a competitive basic research community, and achieved significant levels of industrial self-sufficiency. Hence, while the NICs did not eliminate the gap between domestic and external technological frontiers (i.e., the difference between local and international technological abilities; as is the technology structure in the developed countries; Guimaraes, 1989), they seemingly were able to harness science education to their economic benefit.

Based on these exemplary cases, which show the importance of science education in bringing national economic development, develop-ment programs prescribe two complementary paths to progress in LDCs: (a) the transfer of knowledge and technology from the more developed countries to the less developed countries, and (b) the establishment of a national scientific infrastructure.[32] Such infrastructure is to set the basis for local scientific advances or to enable the appropriate absorption of transferred technology. Science education is an integral part of such a solid scientific infrastructure and its importance as such is widely acknowledged.

Policy documents argue that science education provides the knowledge base for advancing the technical skills of local manpower. Hence, they argue, science education, while being directed at creating a technically skilled labor force, enables *knowledge-based* economic growth. In addition, it is argued that such a national strategy, of building a competitive economy based on knowledge and not on low

[32] The UN and its agencies explicitly advocate both paths to national development. For example, these issues are part of the World Plan for Action for the Application of Science and Technology to Development (UN-ECOSOC resolution 1823(LV), 10 August 1973) and the declaration which followed the 1979 Vienna UN-sponsored conference regarding science and development (regarding conference objectives, see UN-ECOSOC resolution 1155(XLI), 5 August 1966.)

wage assembly, prohibits economic dependency. Moreover, the contribution of science education to economic growth is perceived as effective at all time dimensions: in the immediate term scientifically educated manpower is useful at solving practical problems, and in the long term science education helps establish a local scientific community to set the basis for self-reliant economic development. Rhetorically, these arguments as for the social role of science education are well organized.

Because the social role of science education is intertwined with other social institutions, science education policies call attention to the need for governmental coordination of economic, education and science policies. Governmental control and coordination ensure an efficient allocation of resources and a perfect match of market needs and educational supply.[33] Such governmental involvement is noted as a main factor in the economic success of the NICs (e.g., Guimaraes, 1989) and of Western countries (Hage et al., 1988, for a comparative study of France and the USA). Moreover, centralized control over these policies creates a better, more stable link among the institutions. Specifically, LDCs are encouraged to establish links between university research and industry, so to involve tertiary science education in industrialization and production. For example, the 1974 creation of Daeduk Science Town in South Korea is marked as an exemplary case of the benefits in establishing industrial parks: linking basic scientific research with industrial development, attracting foreign high-tech firms and facilitating effective technology transfer (see Eisemon & Davis, 1991, p. 291). Studies of the role of governments set Japan as an exemplary case of effective, direct government control over the coordination of education, science and economic policies, and note that African countries should follow this example (e.g., Cooper, 1973, p. 7, Eisemon & Davis, 1991).

Although circumstances greatly changed since the massive introduction of science education during the 1960s (see, Lewis 1993), science education policies still advocate the "science education for development" model. Moreover, the re-evaluation and changes in the definition of "national development" lead to a corresponding change in the definitions of education's role (Chabott, 1995) and of science's role in national progress (Clarke 1985, p. 61-2; Drori, 1995; Elzinga & Jamison, 1995). Nevertheless, in both developed and less developed countries science and science education were, and still are, defined predominantly according to their developmental benefits. In other words, the late 1970s shift towards the "basic needs" approach and the importance of "self-reliance", "indigenous R&D", "endogenous development", and "country-", or "culture-specific adjustments", still did not break the link between science education and national progress.

1.4 The Global Expansion of Science Education

Following the international effort to expand mass education during the 1950s, since the 1960s the focus shifted to the promotion of science education. Internationally sponsored, national policies lead science education to take a central role in elementary- and secondary school curricula. While some curricular topics, such as vocational training, are on the decline (e.g., Benavot, 1983), mathematics and

[33] On a more abstract level, the inter-relations among the national policies for science education, technology, or labor further expresses the conceptual link between these institutions.

science education are gaining strength in terms of percent of instructional time. Science and mathematics education became a necessary component in the new, modernized curriculum types in secondary education (Kamens & Benavot, 1991; Meyer, Kamens & Benavot, 1992, Kamens, Meyer & Benavot, 1993.)[34] "[V]irtually all developing countries have mandated some form of mathematics and science instruction in public elementary schools" and lower secondary schools (Benavot, 1992, p. 155).

The subjects of science and mathematics take an average of one third of total official curriculum in both elementary and secondary levels, with but modest variations by the nation's geo-political location (Lee 1990; Kamens & Benavot, 1991; Benavot, 1992; Kamens, Meyer & Benavot, 1993; Lee & Wong, 1990). In spite of the overall strengthening and globalization of science education, the growing emphasis on it is not quite uniform among all nation-states. Overall, LDCs put greater emphasis on science education than do DCs. For example, Latin American and Caribbean countries allocated during the 1980s about 11.2 percent of their primary school curriculum to science subjects, while the same percent in the West is only 6.4 and the worldwide average is 7.5 (Lee & Wong, 1990, p. 25, Table 1). Similarly, Benavot stresses that "less developed nations place *greater* curricular emphasis on elementary science than more developed Organization for Economic Cooperation and Development (OECD) nations" (Benavot, 1992, p. 155).

In addition to the worldwide spread of science education practices, over time science and mathematics instruction increases its relative share of education. Nation-states increase enrollments in science and engineering in their higher education sector (Ramirez & Lee 1990), and all devote a greater average share of instruction time to mathematics and science than they used to in the past (Kamens & Benavot 1991, Benavot 1992:156). Few countries are exceptions to this trend; Asian and Latin American countries shows a slight average reduction in the share of science and mathematics instruction time in their secondary schools (Benavot 1992:155-156.) Overall, this trend of growth in attention to mathematics and science instruction is a part of the general trend of establishing science instruction as an integral part of mass schooling, but it is particularly noticeable in the present period (Kamens & Benavot 1991).

In summary, science education has greatly intensified. There is an overall expansion in curricular relative share of mathematics and science instruction, policy statements increasingly emphasize the importance of mathematics and science instruction to national development, and there is an increase in science enrollments. In other words, the expansion of science education is actualized on several levels: (a) in an increase in the volume of science education (i.e., science and mathematics education gets greater attention in primary and secondary schooling in terms of instructional time, teacher training, resources etc.), and (b) in the introduction of science education practices to an *increasing number of nation-states* (i.e., more nation-states introduce programs for science and mathematics education.) In general, the expansion of science education transcends developmental barriers and geo-political divisions.

[34] For a thorough review and analysis of the historical processes of making science- and mathematics-related subjects into core elements in school curriculum since the 1800s, see Kamens & Benavot 1991.

Expansion and globalization trends in science education lead to standardization of the field and to isomorphism[35]. The variance among nation-states in their commitment to science education is small and declining (Kamens & Benavot 1991.) The worldwide standardization of science education – in curricular emphasis and expansion trend – suggests that it is not a reflection of local and national interests, but rather of trends in the world polity. In other words, the homogeneity of, for example, mathematics and science curriculum implies that the theme of "science for development" and the related focus of science education are drawn from a common, global cultural source (i.e., the world polity) rather than being simultaneously initiated by particularist sources.[36]

To conclude, the conceptual model of "science education for development" sets the basis for national policies of science and of education, for international aid efforts, and for trans-national cooperation. Much of the empirical evidence for the validity of this policy model is exemplary -- the NICs and the history of the West. Current social science comparative research provides, at best, conflicting evidence for its validity. Part II of this chapter examines the empirical evidence to refute the existence of a positive link between science education and national development and the arguments that support these findings.

Part II: A Critical Assessment of Science Education for Development

Science education policies were, and still are, under attack. When these policies were initially formulated and implemented during the 1960s, the arguments concerning the disabling consequences of education, science, technology and international economic policies in LDCs were already fully developed. Dependency theory writers, most notably Emanuel Wallerstein, wrote extensively on the role of these institutions in the perpetual underdevelopment of Third World countries. However, while elaborating on the subjects of science (e.g., Sagasti 1973) and of education (e.g., Mazrui 1975), very few analyses focus exclusively on science education as a mechanism of control. The criticism of science education for economic development can be divided into three categories: (a) a challenge of the findings of positive connection between science education and economic development, (b) an assessment of the mechanisms that restrain such a causal link, and (c) a discussion of the ways by which science education is involved in promoting and expanding underdevelopment.

2.1 Science Education and Economic Growth: An Observed Link?

Recent empirical studies, while testing the causal link between science education and national economic development, reveal findings that contradict our everyday notion

---

[35] Isomorphism is defined as structural similarities and copying mechanisms within an organizational field, or within an institutional environment. In the context of globalization processes, isomorphism means that countries tend to format their national structures and institutions according to existing and legitimized national formats in their surrounding world polity. For the classic discussion of isomorphism and its mechanisms, see DiMaggio & Powell 1983; for an application of the concept of isomorphism to the globalization of science, see Shenhav & Kamens 1991.

[36] For additional institutionalist arguments regarding the global expansion of science education, see Kamens & Benavot 1991, Benavot et Al. 1991, Benavot 1992, Meyer, Kamens & Benavot 1992. Also see Part III.

of science education and the predictions made by the model of "science education for development." These studies show that official emphasis on mathematics and science studies in primary education (Kamens & Benavot, 1991, p. 168-169) and the share of instruction time devoted to mathematics, natural sciences or social sciences (Benavot et al., 1991, p. 95) have no effect on economic conditions. More specifically, Kamens & Benavot (1991, p. 166) summarize that "it does not appear that official attention given to mathematics and science instruction in primary education is directly related to key indicators of socio-economic development, economic dependence, or world system position."

Case studies of similar effects support these findings and show that national commitment to science education is adversely related to economic development. For example, college degrees awarded in science-related fields in the USA have no effect on productivity. "We find no evidence that degrees awarded in math, engineering or physical sciences - the fields in which students are expected to learn the skills that are more directly relevant to occupational demands – have any positive effects on changes in labor productivity" (Walters & O'Connell, 1990, p. 18; see Walters, 1989).[37] In summary, such studies show that the link between science education and economic progress is more complex than previously thought.

Moreover, a careful examination of the phenomenal case of economic development in the NICs shows that there is no corresponding high degree of relative emphasis on science education. In other words, the NICs enjoy high rates of economic growth *in spite* of the fact that they trail other regions in the relative share of curriculum devoted to science subjects. This is while sub-Saharan African nations, which suffer economic stagnation, devote greater shares of official curriculum to science and mathematics (see, Lee & Wong, 1990, p. 21-22; Kamens & Benavot, 1991, p. 164-165).[38] This comparison between the NICs and sub-Saharan countries on the matter of their time-share of science education compared with their economic growth doubts the causal link between science education and economic prosperity. As Ajeyalemi (1990, p. 121) summarizes his study of seven sub-African countries: "Overall, formal [science and; GD] technology education in Africa is yet to make the needed impact on development." The fact that struggling LDCs, such as sub-Saharan African nations, are devoting a high relative share of their primary curriculum to science education is an example of *over-education*, i.e. over-emphasizing science education relative to other topics and in comparison with the proper balance needed to produce best results. In terms of cost efficiency for governmental expenditure, such over-education is frequently regarded as a waste of increasingly scarce governmental resources.

These later empirical tests of the "science education for development" model

---

[37] These studies, showing no relationship or negative relationship between science education and economic development, are supported by similar analyses regarding the relationships between mass education or general scientific activity and economic development. Cross-national studies found that tertiary education (Meyer et al., 1979)and scientific paper publication (Shenhav & Drori 1988, Shenhav & Kamens 1991) have no effect on economic growth. Case studies show that in the USA college enrollments between 1933 and 1969 have no lagged effect on economic output (Walters & Rubinson 1983.)

[38] For a refutation of the myth that emphasis on, and attainment of, education established the unique pattern of economic development in Asian countries, see Baker and Holsinger 1996.

challenge the modernization theory's assumption that the path for national development is universal. In other words, these studies show that "receipts" for economic development, such as through science education and scientific and technological infrastructures, do not produce globally uniform results. Rather, the inter-relations between science education and the economy differ greatly between DCs and LDCs, among different world regions, among the nations which are a part of the big and varied group of LDCs, and even between regions, social groups, and cultural zones within each nation (e.g., Grindle & Thomas, 1991, p. 97-99; Mordi, 1993.) Even nations that seemingly share present conditions are divided by their histories, such as colonial experiences, and, thus, are divided by paths of progression. In summary, empirical evidence reveals that the causal relations between science education and economic development are more complex and diverse than the science education policies of the 1960s and 1970s anticipated.

2.2 What Social Mechanisms Stand in the Way of Success?
Several social mechanisms mediate between science education and the economy and are noted as restraining their effective positive relations. Elementary as it may sound, the lack of data on science education, scientific prospectives for graduates, scientifically skilled manpower, and governmental investment in R&D makes any analysis and re-evaluation practically impossible (Lewis 1993, p. 5; Eisemon & Davis; 1991, p. 282). In most countries, and especially in LDCs, such information is either not collected at all or not gathered on a regular basis. Such lack of information and data is a result of the administrative weakness of the LDCs' governments (see, Eisemon & Davis, 1991, p. 282; Shrum & Shenhav, 1995, p. 25). This administrative failure is also exhibited in the inability to coordinate education, science, and economic policies and to device long-term goals.

Another administrative problem in LDCs that has grave consequences for the effectiveness of science education is the issue of resources and funding. Declining available resources result in lower levels of science teaching and materials available for effective teaching (Lewis, 1993, p. 4). Hence, poor countries are less likely to offer quality science education (in terms of teachers' training, available school laboratories and other material resources such as maps and books) as part of primary- and secondary school curriculum. Lee & Wong (1990) show that (a) "the likelihood of a country offering a science track was positively related to the government revenue as a percentage of gross domestic product" (p. 27), and that "countries that were highly dependent on foreign investment were less likely to place much emphasis on the teaching of science in the upper secondary level, either in the science tracks or in the comprehensive tracks" (p. 29).

In addition, the process of loose coupling[39] inhibits any linear, causal effect among the different institutions in the scheme of "science education for development." In other words, science- and education related state institutions are not tightly organized, thus their policies, actions, and resources are poorly coordinated. The loose coupling between science policy and science practice in LDCs (Ramirez & Drori, 1992) hold back any implementation of science-related

[39] Loose coupling is defined as weak institutional, organizational or operational connections among organizational units.

guidelines. Science and technology are also not linearly connected, but rather each exhibits an independent and unique effect on the economy in LDCs (Madvitz, 1985; Drori, 1993.) The relations between science education and technology education is practically an un-explored field (Gilbert, 1992.) The expansion and size of the scientific community in LDCs is also not related to scientific productivity (e.g., Eisemon & Davis, 1991, p. 280, regarding Asian and African universities) or to the productivity of the labor force. In summary, in LDCs the different institutions that are epistemically linked (in a causal manner) in the conceptual model of "science education for development" – science education, scientifically educated and skilled manpower, scientific and technological productivity, and, finally, economic benefits – demonstrate weak ties among themselves and do not seem to operate in a linear fashion, as expected by experts. Thus, in spite of programs for the promotions of science education and general science, scientific infrastructure is non-existent in LDCs (e.g., Zymelman, 1990, for a study of sub-Saharan Africa.)[40]

To summarize, I described but a few mechanisms that halt the causal effect of science education on economic development, especially in LDCs. These mechanisms concern, to a greater extent, governmental and organizational procedures. Nevertheless, the ineffectiveness of science education in terms of economic progress is also due to, and reflective of, cultural barriers.[41] Such barriers impede the transfer of knowledge – knowledge that was created in a particular cultural context and thus is value- and symbol-laden – to another cultural sphere. While some barriers are officially set as cultural "defense" mechanisms (such as the government rules in some Islamic countries against the importation and for censorship of cultural material, including science information), most nation-states are infused with informal mechanisms. These cultural barriers further halt the expect results from global and national efforts towards scientization.

2.3 Science Education and Underdevelopment

Dependency theorists[42] view science education, especially in LDCs, as carrying a hidden agenda of perpetual underdevelopment. They view the long-term objectives of science education policies as long-term hopes with adverse short-term consequences. The current format of science education – its content and relative emphasis – maintain, and further contribute to, the cultural and technological domination of the West over Third World Countries, and, thus, to the primary form of domination – economic domination. Hence, the lack of positive effect of science education on the economy of LDCs is because this education is devised to expand the basis for Western capitalism by shaping the scientific and technological future of LDCs to match the economic need of the Core.

How does science education contribute and expand underdevelop-ment? Science education is criticized as providing *mis*-education. In other words, it provides education that is not applicable to local needs and it channels the labor forces towards a narrower, rather than wider, range of occupational opportunities. Since the quality of science education in LDCs is poor (in terms of resources and,

[40] For a review of the state of science in LDCs, see Shrum & Shenhav 1995.
[41] For further insight on the issue of cultural barriers see Chapters 7, 8, and 9.
[42] Refer back to Cobern's discussion of cultural deficit and dependency in Chapter 1.

thus, of quality), it provides only low-level technical skills. Such poor skills are applicable only for the absorption of transferred technology, rather than for local innovation. And, reliance on transferred technology in LDCs, in the absence of proper scientific and technological infrastructures, leads to technological dependence on the economy of DCs. In summary, science education, in its current poor state in LDCs, re-establishes the economic dependency of LDCs on DCs through the expansion of technological- and labor market dependencies.

In addition to science education not being applicable to local levels of technological and industrial sophistication, science education is also not directed at being applicable to local needs. Science, as well as science education, tends to follow the research trends and academic "fashions" set in the scientific core[43]. There is a worldwide trend toward the standardization and homogeneity of science curriculum in primary and secondary schools (McEneaney, 1995.) Such trends affect the number and type of scientific subjects that are taught in schools or the type of pedagogical devices used, etc. For example, increasingly national science curricula include environmental studies and exclude the previously taught subject of anatomy. Similar changes occur regarding pedagogy: in science textbooks worldwide there is a trend toward experimentation and other activity-based learning, and away from learning by rote.

Such trends towards global homogeneity are somewhat contained by efforts to "nationalize" science curriculum. For example, illustrations in primary level science textbook in Ghana include black-skinned people, and Hungarian secondary level physics textbooks include a chapter about renowned Hungarian physicists (McEneaney, 1995.) Such efforts to "localize" science curriculum do not, however, contain the problem of alienation of local educated people from their social context. In stead of solving practical problems and responding to "burning" local concerns, scientists tend to focus on research questions that are aloof to their local context. As S. Kionga-Kamanu, of the University of Nairobi, says: "There is little glory or recognition for improving a 'jiko,'[44] for instance, compared to work in the area of nuclear physics. It is obvious, however, which of these efforts has more direct applicability and potential use for the people in Kenya" (quoted in Bruchhaus, 1985, p. 26). It is science education, in its Westernized form and aspirations, that sets a gap between Third World scientists and their society and places them in face of this dilemma.

The effectiveness of science education programs worldwide is judged by the standard achievement tests in mathematics and the sciences, such as those devised by Educational Testing Service (ETS) for The International Assessment of Educational Progress (IAEP)[45] or the Third International Mathematics and Science Study (TIMSS). While these tests are administered in a few Third World countries, similar tests are given in LDCs by other agencies (e.g., Mordi, 1993). In order for students to succeed in such tests, if for no other reason, mathematics and science curriculum

---

[43] See, for example, Benavot et al., (1991) for a cross-national study and Lee (1992) for a study of Malaysia.

[44] Jiko is the traditional and still popular cooking stove in Kenya.

[45] For examples of the publications of the cross-national tests, comparative results and analyses, see ETS-IAEP's Learning Science and Learning Mathematics (Lapointe, et al., 1992.)

is also standardized to follow the worldwide, core-defined definition of science education. In other words, since the tests are designed to monitor specific knowledge and skills, school curriculum offers the same topics in preparation of the tests. Moreover, science education programs in LDCs are created to enable pupils to be accepted to undergraduate and graduate studies abroad, where science studies are closer to the "frontier of knowledge." Overall, science education in LDCs is formulated in light of administrative and academic requirements in the Core countries. This process further alienates science education from local needs and national context.

When considering governmental resources, science education is criticized as a diversion of critical, scarce funds from their use to solve immediate problems. In other words, an increasing number of governments in Third World countries are funding science education, while still troubled by the question of how to distribute their limited resources. Again, the balance between science education and long-term plans, on the one hand, and short-term needs, on the other, is brought into question. This dilemma is most acute in LDCs struggling with widespread hunger, political instability and violence, and natural disasters.

Available resources for science education, however, do not insure that science education is widely implemented and evenly distributed. Rather, countries with strong state fiscal power and strong state control over education, such as Malaysia, Singapore, and Hong Kong, are more likely to establish elitist science education by implementing science specialization in their upper secondary schools (Lee & Wong, 1990, p. 29). Overall, in LDCs proper conditions for science education are available only to a selected social group (Lewis, 1993, p. 4). In summary, a greater amount of national resources does not necessarily lead to redistribution of the wealth to all walks of society. Rather, when such resources are available, they are being used by, and to the benefit of, the local elite, who is positioned as the mediator in the process of underdevelopment between the West and local proletariat.

Most importantly in the process of exploitation and underdevelopment is cultural dependency. The idea of development, and the institutions whose justification rest on it such as science education, are Western pathologies, which set taxonomies, divisions and, thus, control mechanisms.[46] Having its roots in colonial history, modern science is alienated from non-Western contexts (e.g., Goonatilake, 1984; Nandy, 1988). Its introduction through science education means indoctrination into Western logic and procedures, and it furthers the alienation of the educated few from their local society.

In summary, dependency theory sets science education as sustaining dependency in LDCs and in the weaker sectors of advanced nation-states. It argues that science education is an un-necessary commitment of scarce funds to an institution that serves the exploitative interests of the powerful core countries and of their capitalist needs. Because science education is designed to satisfy such needs, it is detached from local context and local problems. In LDCs, although science education is increasingly institutionalized and although an increasing share of

[46] Representing radical dependency arguments, Alvares (1992, p. 94) states, most bluntly, that "[d]evelopment is best understood as a project through which an aggressive, expanding class seeks to expand its control and use of other people's resources and to neutralize any opposition to such programs."

instruction time is devoted to it, it is still an elitist sector of education. It is elitist in its aspirations and in the fact that it reaches only a small and privileged sector of society in the developing countries.

2.4 Summary: Realist Perspectives on Science Education for Development

Policy directives emphasize science and mathematics instruction as having measurable economic effects. This emphasis sets the background for worldwide calls for educational reforms. It is assumed that such reforms, by improving instruction, providing teachers' training, and re-structuring the organizational format of education, will enhance the effectiveness of science education, and, in turn, will result in greater economic benefits. However, these policy assumptions are rarely empirically tested. Very few cross-national studies focus on the effects of science and mathematics education on national economic development. Moreover, the evidence provided by these studies is inconclusive. Hence, there is a need for further empirical testing of the causal, hierarchical connection, assumed in the conceptual model of "science education for development."

Moreover, the lack of conclusive empirical evidence regarding this connection prohibits any adjudication between, or at least any evaluation of the validity of, the competing liberal- and Marxist-based theories of national development, i.e., the modernization and dependency theories, respectively. Regarding the social role of science education, these theories diverge on their definition of the motives and processes (functional innovations versus imperialist impositions) which lead to the institutionalization of science education, in their conceptualization of the type of benefits in establishing science education, in pointing to the beneficiaries from the institutionalization of science education, in identifying the interests (national versus foreign, mass versus elite) that are served by science education, and in description of the mechanisms that connect science education with national economic development. Advocates of each of these theories cite empirical evidence to support their arguments and consider the contradicting evidence to be a methodological, or statistical, artifact. However, the value-laden nature of each theory makes them somewhat incommensurable (see Epstein, 1983). These debates place the notion of national development in a dialectic position, where means for progress that are emphasized by modernization theory, such as science education, are turned into means of marginalization, underdevelopment and exploitation, as emphasized by dependency theory (see Apter, 1987). In this sense, the debate between the two notions of what is national development has reached an impasse. Accordingly, when science education policies are revealed as theory-laden and empirical evidence provides merely inconclusive evidence, national policies for science education should be devised and implemented with greater caution, especially in LDCs. Moreover, our conceptualization of science education may have influenced us in searching for policy direction. In other words, we should alter our definition of science education. As Medvitz (1985) argues: "[i]n order for school science to have a desirable effect on technological and economic development, it is important to recognize science as a cultural form." As such, science is not applicable to all cultural contexts. Science education policy should, thus, be cautiously implemented, since we do not hold sufficient knowledge about the ways by which cultural forms affect national development.

Over and above the policy implications, the theoretical debate reveals an additional important point. In spite of the many diverging points between these theories, they also share several core elements. For example, both theories of national development argue for a global perspective when considering issues that are seemingly intra-national, they offer a structural perspective on social processes, and both theories define social needs and device the ways to satisfy them. Most importantly to this discussion, however, is their mutual embrace of science as a means for national development. Both theories share a functionalist, utilitarian, and *realist* epistemological attitude towards science education. In other words, both theories argue that science education *does* produce results, and their debate is over *which* science produces best results. The arguments set by dependency theory criticize international and national science education policies and their modernization-laden arguments in the name of *real* science.

Much in response to these realist-consequentialist perspectives, I argue in Part III that science education is a socially constructed institution. In other words, it is an organizational *manifestation of cultural definitions* of its social role, of its organizational format and of its patterns and procedures. This re-definition of science education is not realized by both modernization and dependency theories. Whatever the stand in the "good science" - "bad science" debate, science education is still conceptualized in both theories in a utilitarian, or consequantialist, manner. Both dependency and modernization, hence, reduce the notion of science education to uni-dimensional, development-dependent definition that is formulated in the "science education for development" conceptual model. Neither perspective – modernization nor dependency theories – comes to terms with the symbolic character of the worldwide institutionalization of science education. The existence of the institution of science education is not an objective "reality", but rather an outcome of a particular cultural context; hence, science education is an institution that is "real" only in its consequences. By taking this epistemological stand, I break away from the theoretical posture presented in Part II.

Part III: Science Education, National Development, and the Discourse of Global Policy

3.1 Science Education Policy as a Narrative

Dissatisfied with the consequencialist emphasis in both theories of national development – the liberal modernization theory and the radical dependency theory – I turn, for both stimulation and re-direction, to discourse analysis[47] and to the institutional perspective.[48] I suggest that we re-think and re-conceptualize a discursive analysis and institutional analysis. I suggest that we re-think and re-conceptualize the current policies for science education: these policy texts are narratives of modernity that were globalized and legitimized through organizational procedures. In other words, such policies and the organizational structures that are created upon their recommendations are, essentially, a reflection of global,

[47] Along the epistemic lines of French structuralism.

[48] The comparative variant of institutional theory, also labeled "world polity perspective", is best explained in Meyer, Boli & Thomas (1987). Also see Zucker (1987) and Scott (1987) for reviews of the emergence of, and themes in, institutional theory.

legitimized, social myths.[49] Regarding our discussion, the global myth concerns the social role of science education and the content of this global myth is the conceptual model of "science education for development."

What is a global myth? And, how is the model of "science education for development" a global myth? A global myth is a world value; in other words, it is a desired social merit of modern nation-states, or a characteristic that is perceived as a highly valued national trait. The policy model of "science education for development" holds the main defining features of a global myth.

- It is global; in other words, science education – as a concept, a policy, and a governmental action – is institutionalized worldwide. As mentioned earlier, there is a consistent trend towards further globalization of this policy model.
- It is legitimized; in other words, all nation-states consider the social role of science education to be a valid and core concern of modern society, and the national preoccupation with science education is considered proper. Such legitimacy greatly relies upon the linkage among science, education and national progress, as progress is a core element in modern Western thought.
- Like other global myths related to national society,[50] the "science education for development" model is highly scientized; in other words, the modeling of the social role of science education is formatted according to scientific definitions and procedures of logic and testing. Hence, science education programs are structured in a logical manner and scientifically analyzed, measured, and tested.
- Most central to my argument, this model is a common article of faith, a taken-for-granted, or a social convention. The myth of "science education for development" as a whole, unlike some of its components (such as, specific science education techniques and materials or particular teachers' training programs) is rarely questioned and re-evaluated. In other words, there is a shared belief that science education is effective, necessary, and produces results, while there is little proof for the empirical validity of such a common belief.[51] Science education is among the institutions that are defined as desired components of modern nation-statehood (and a required component of modernity). Above all other global myths, science education carries the glory of the totalizing effect of science, technology and education in the shaping of modern society. With them, science education also shares its stature as a core element of modernity and an important factor in bringing "salvation" (see Rosenberg, 1976; Midgley, 1992).

3.2 The Historical Processes linking between Science and Development

How and when was the myth of "science education for development" constructed? When and why was science causally linked with the concept of national development? "It is a curious matter that modern education [and also science education, which is a core component of it; GD] – so dramatically tied to (and, indeed, a component of) the functional theories of modern society (including left-

---

[49] I conceptualize "myth" in the anthropological sense. In other words, myth is a taken-for-granted, non-contested, "sacred" cultural element.

[50] Such as (a) the state as the guardian of the nation, (b) the individual as a social "unit", (c) the state and the nation as aggregates of individuals, (d) childhood socialization and the life-course as the defining processes, and (e) progress as a necessary process and desired value. See, Thomas et al., 1987.

[51] Policy makers see empirical evidence for this assumption in the statistical findings that science educated people have higher incomes. Based on it, they falsely make inferences to the nation-state level. Such inference is based on another global myth - that a nation-state is an aggregate of individuals. In summary, the lack of direct evidence leads to inference, which is itself based on assumptions and articles of faith.

wing variants) – is probably more a creature of these theories than of the reality they supposedly portray" (Kamens, Meyer & Benavot, 1993, p. 14). Indeed, the conceptual model of "science education for development" is a construction of a particular historical moment, in a particular geo-political location, and, most importantly, in their particular cultural setting. Whereas science and education are commonly regarded as *intrinsically* being linked with social benefits and, thus, *by definition* carrying a teleological tone, I argue that such definition evolved in the cultural environment of nineteenth century Europe.

Science and education are regarded as being beneficial to society and the definition of their social role rests on this justification. In other words, the social understanding of science and education is essentially teleological, and modern science and education are regarded as being linked to, and defined by, their utility. Moreover, the current historiography of science, for example, assumes that modern science evolved as teleological during its emergence in the seventeenth century.[52] I argue that during the seventeenth century the Western theory of utility was not yet developed and consolidated sufficiently to affect the conceptualization, and later the operation, of social institutions. I view the social atmosphere of the nineteenth century – with the processes of secularization, democratization, and professionalization – as setting the social environment and enabling the construction of the link among science, its utilitarian qualities and the nation-state. Moreover, this social context fostered the retrospective reconstruction of science in a teleological light. In this context, the scientists of the late nineteenth century and early twentieth century, in their search for greater legitimacy and for a stable definition of their profession, created this conceptual link between science and utility and re-told the story of the seventeenth century roots of modern science.[53]

Following Bourdieu (e.g., 1990, p. 185), I view social truths as rooted in the historical, social and cultural context of the period during which they were defined. Consequently, I emphasize that the perception of science education as resulting in social advantages is a social construction. From this perspective, science education is posed against a socially constructed need – economic development. By focusing on the historiography of modern science and on the social context in which it was written, I argue that the popular policy model of "science education for development" is a construct of social theorizing, and of the professional motives and cultural context in which such theorizing is done. In this social context, where utility and nations-statehood were idealized, science and education were legitimized *because* they were defined as enabling the achievement of such ideals.

Moreover, the current trends of expansion of science education are enabled only due to the fact that the *value* of science education has already been established. In other words, only once the social role of science education is institutionalized and legitimated as proper and necessary, can the field of science education expand. It is this particular definition of the social role of science education that allows for the

---

[52] These historiography rests an reading of seventeenth century writings, such as Francis Bacon's *New Atlantis* (1627), where the practical contributions of science are described: improved orchards, improved breeds of animals, and improved medications.

[53] For a broader description of my thesis regarding the role of the nineteenth century Western environment, and, mainly of the professionalization of science, in the retrospective reconstruction of the science-utility link, see Drori (1994).

great investments in science education. The definition of the social role, or the designated value of this social role, is exemplified in the "science education for development" model. This model is, however, a general and widespread article of collective faith, rather than an observable reality. In other words, the science education component in national development is a myth, rather than a proven agent for national progress.

3.3 Perception and Globalization

After being institutionalized and legitimized in the West, this conceptual model, that established a causal link between science and economic progress, was diffused worldwide. How is this myth globalized? How does this notion become accepted worldwide? The evidence that science education is expanding worldwide and that there are very few particularist varieties to science education programs, suggest that all policies of science education draw from a single source.[54] This source is the world polity, which is a web of globalized myths and their organizational carriers. In other words, this "umbrella" of international culture includes the hegemonic ideals and the organizational network that sponsors these ideals and transfers them to all member nation-states. Ideals, which are *perceived* as being successful, set as exemplars and copied from the core through processes of isomorphism. Once economic development was defined as a social desideratum, the search for an answer for this "national problem" centered around (among other things such as entrepreneurship and investment) education, science, and specifically science education. The economic success of Western countries and later of the NICs was *perceived* as depending on levels of literacy and scientific achievement. Consequently, weaker countries adopted the seemingly successful model of "science education for development." Even once some critical arguments were raised against its implementation, core countries still adhere to this model. For example, American education policy is guided by the notion that American children are lagging behind South-East Asian children in standardized, comparative science tests, and that the industrial, economic, and defense future of the U.S. depends on such matters.

Models are copied only once, and for as long as they are perceived as successful. Since the 1950s mass education was perceived as the key factor for achieving economic development. Hence, development policies emphasized the importance of institutionalizing mass education in every country. Since the 1970s, however, the focus shifted from general literacy to science education, and scientific literacy is currently defined as being at the core of the causal link between mass education and economic development. Consequently, science education is currently perceived as the successful factor. This 1970s shift is much a reaction to the 1960s notions of mass education that emphasized *basic* education or vocational training. This basic education approach was heavily criticized as establishing second rate education, which results in a dependent economy.[55] These thematic changes in policy later translated into actual science education programs. Science education, much like general education programs, is hence "extremely sensitive and responsive

---

[54] For further examples of curricular similarity among different nation-states, see Benavot, et al., (1991), Meyer, Kamens & Benavot, (1992). For a case study of global influences on school curriculum in Malaysia, see Lee (1992).

[55] For review and analysis of thematic changes in education policy worldwide, see Chabbott (1995).

to world scientific and professional ideals and ideologies" (Hufner, Meyer & Naumann, 1987, p. 207).

This ideal of national progress is heavily promoted by both science- and development-related organizations. As mentioned earlier, international and national organizations adopted the model of "science education for development" and became its carriers worldwide. UNESCO, for example, organizes international and regional conferences to promote this idea, it researches the effect of science education on economic development both cross-nationally and in specific countries, and it supports (in terms of financing and advising) the establishment of national science education programs.[56] In this sense, the global myth of "science education for development," while being a social construction, carries "real" implications. Once acknowledged and institutionalized as a social desideratum and once it is implemented into policies and action, the myth changes the organizational structure of all participating nation-states. In other words, the global myth of "science education for development" leads to the establishment of science tracks in primary- and secondary schools, to a greater emphasis on science and mathematics in school curriculum, to the development of school laboratories, and to a greater investment in, and attention by politicians to, the issue of science education. Such "real" consequences may even be economic development. Such economic consequences should, nevertheless, be conceptually re-framed; as Benavot (1992, p. 173) argues, "the economic effect of science education may have more to do with 'hidden' cultural rules, orientations, and worldviews being transmitted than the specific scientific content being taught." In summary, the local organizational *map* and local social procedures are, hence, imprinted with the impression of the world polity – its myths, its institutional ties, and its organizational form.

Part IV: Concluding Remarks

Science education is a popular topic in both the education and the development agendas. In a world where economic progress is a desired social goal and, most often, a national problem, mathematics and science instruction is currently emphasized as a possible solution. The causal link between science instruction and national progress is, however, empirically problematic and conceptually misleading. In other words, from a realist standpoint, there is conflicting evidence for the validity of this causal connection, and from an epistemic standpoint, this policy model is celebrated while disregarding its socially constructed nature.

In this chapter I demonstrated how, while the realist perspectives continue their arguments whether science education does or does not produce the desired social goal of economic development, their discussions have reached a theoretical impasse and an empirical deadlock. World polity perspective, however, while investigating the agenda of science education as a discursive narrative, allows for a variety of stimulating research directions. Such possible research directions can be divided into two categories: First, comparative studies of the content and trends in the discourse of science education, and second, comparative studies of the effects of specific discursive regimes in the field of science education.

[56] For a study of international organizations, and specifically UNESCO, as "teachers of norms", see Finnemore (1990 & 1991).

First, the nature of the discourse of science education needs to be further explored. Such investigation can examine the content of the discourse and describe the nature of the narrative. Science textbooks, teachers' training material, pedagogical techniques and policy statements (both of international agencies and of local governments) may serve as exemplars of the discourse. In addition, an examination of such discursive themes over time may reveal the changes in the content of the narrative and the trends in its development. Examining such trends in relation to global events and processes may lead to a better understanding of the factors that are involved in the development of the policy model of "science education for development." Finally, there is a need for further examination of the historical process, by which the current discourse of "science education for development" was formulated. Such investigation will reveal the social groups that were, and maybe still are, involved in the construction of this discourse, and who serve as its carriers, their power relations, their formal and informal aims etc. In summary, the examination of a specific discursive regime in science education can investigate the content of the narrative, the trends or changes in the narrative, and the historical process of the construction of the narrative.

A second direction for further investigation is the research of the effects of such discursive regimes and of their trends of change on society. Being analyzed as a discourse, the policy model of "science education for development" is "mere words", which reflect a particular social context. Nevertheless, such "words" have great simultaneous impact on their social environment. In other words, while the discourse is a social construct, it is real in its consequences. Policy directives guide action and form organizations; education programs lead to the writing of textbooks and to the choice of pedagogical devices; and, such textbooks and pedagogy teach children specific facts and construct their world and their interpretive framework for it. These, and other, consequences of the current policy models shape our social world. Yet, such consequences of science education policies are not explored in current research. In conclusion, the institutionalist, comparative and discursive analyses of science education enable new venues for innovative social research.

References

Ajeyalemi, D. (1990). Science and technology education in Africa: a comparative analysis and future prospects, in D. Ajeyalemi, (ed.), *Science and Technology Education in Africa: Focus on Seven Sub-Saharan Countries*, Lagos, Nigeria: University of Lagos Press, p. 114-127.

Altbach, P. G. (1989). Higher education and scientific development: the promise of the newly industrialized countries, in P. G. Altbach, et al., (eds.), *Scientific Development and Higher Education: The case of Newly Industrialized Countries*, NY: Praeger, p. 3-29.

Alvares, C. (1992). *Science, development, and violence: the revolt against modernity*. Delhi: Oxford University Press.

Anderson, C., Bowman, A. & Bowman, M. (eds.) (1965). *Education and economic development*. Chicago: Aldine Publishing.

Apter, D. E. (1987). *Rethinking development: modernization, dependency, and postmodern politics*. Newbury Park, CA: Sage Publications.

Baker, D. P. & Holsinger, D. B. (1996). Human capital formation and school expansion in Asia, *International Journal of Comparative Sociology*, 37(1-2): 159-173.

Benavot, A. (1983). The rise and decline of vocational education, *Sociology of Education*, *56*(2): 63-76.

Benavot, A. (1992). Curricular content, educational expansion, and economic growth, *Comparative Education Review*, *36*(2): 150-74.

Benavot, A., Cha, Y., Kamens, D., Meyer, J. W. & Wong, S. (1991). Knowledge for the masses: world

models and national curricula, 1920-1986, *American Sociological Review*, *56*(1):85-100.
Blute, M. (1972). The growth of science and economic development, *American Sociological Review*, *37*(4): 455-464.
Bourdieu, P. (1990). *In other words: essays towards a reflexive sociology* (translated by Matthew Adamson), Stanford, CA: Stanford University Press.
Bowman, M. & Anderson, A. C. (1963). Concerning the role of education in development. In C. Geertz (ed.), *Old Societies and New States*, Berkeley, CA: University of California Press.
Bruchhaus, E. (1985). Improved cooking stoves: miracle weapon in the fight against the desert? *Development and Cooperation*, 1/85:24-26.
Chabbott, C. (1995). Constructing educational development: the role of international development organizations. Paper presented in *Nordic Association for the Study of Education in Developing Countries Annual Conference*, As, Norway.
Clarke, R. (985) *Science and technology in world development*. Oxford, NY: Oxford University Press/UNESCO.
Colcough, C. (1982). The impact of primary schooling on economic development, *World Development*, *10*:166-185.
Cooper, C. (1973). Science, technology and production in the underdeveloped countries: an introduction. In C. Cooper (ed.), *Science, Technology and Development*, London: Frank Cass, p.1-18.
DiMaggio, P. & Powell, W. W. (1983). The iron cage revisited: institutional isomorphism and collective rationality in organizational fields, *American Sociological Review*, *48*(2): 147-160.
Drori, G. (1993). The relationships between science, technology, and the economy in lesser developed countries, *Social Studies of Science*, *23*(1): 201-215.
Drori, G. (1994). Tracing the intellectual origins of current themes in science policy: The modern state, utilitarianism, and science historiography. Paper presented in American Sociological Association Annual Meeting, Los Angeles, California.
Drori, G. (1995). Science-for-Development: Discursive themes and organizational expansion in national science policies. Paper presented in Society for Social Studies of Science Annual Meeting, Charlottesville, Virginia.
Eisemon, T. O. & Davis, C. H. (1991). University research and the development of scientific capacity in Sub-Saharan Africa and Asia In Altbach, Philip J. (ed.), *International Higher Education - An Encyclopedia*, 1:275-295.
Elzinga, A. & Jamison, A. (1995). Changing policy agendas in science and technology. In J. Sheila, G. Markle, J. Petersen & T. Pinch (eds.), *Handbook of Science, Technology, and Society*, Newbury Park: Sage, pp.572-597.
Epstein, E. H. (1986). Currents left and right: Ideology in comparative education. In G. Kelly & P. Altbach (eds.), *New Approaches to Comparative Education*, Chicago: University of Chicago Press, pp. 233-259.
Ferguson, J. (1990). *The Anti-politics machine: Development, depoliticization, and bureaucratic power in Lesotho*, Cambridge, UK: Cambridge University Press.
Fiala, R. & Gordon-Lanford, A. (1987). Educational ideology and the world educational revolution, 1950-1970, *Comparative Education Review*, *31*(3):315-32.
Finnamore, M. (1990). International organizations as teachers of norms: UNESCO and science policy Paper presented in American Political Science Association Annual Meeting, San Francisco, California.
Finnemore, M. (1991). *Science, the state, and international society*. Unpublished Ph.D. Dissertation, Stanford University - Department of Political Science.
Gilbert, J. K. (1992). The interface between science education and technology education, *International Journal of Science Education*, *14*(5): 563-578.
Golden, H. H. (1955). Literacy and social change in underdeveloped countries, *Rural Sociology*, *20*:1-7.
Goonatilake, S. (1984). *Aborted discovery: Science and creativity in the Third World*, London: Zed Press.
Grindle, M. S. & Thomas, J. W. (1991). *Public choices and policy change: The political economy of reform in developing countries*, Baltimore: Johns Hopkins University Press.
Griliches, Z. (1987). R&D and productivity: Measurement issues and economic results, *Science*, *237*:31-35.
Guimaraes, F. (1989). Technology policy in newly industrial countries: A Brazilian perspective, *Science and Public Policy*, *16*(3): 167-175.
Hage, J., Garnier, M. & Fuller, B. (1988). The active state, investment in human capital, and economic growth: France 1825-1975, *American Sociological Review*, *53*(6): 824-837.

Harbison, F. & Myers, C. A. (1964). *Economics, manpower, and economic growth*, N.Y.: McGraw-Hill.

Heyneman, S. & Siev-White, D. (1985). *The quality of education and economic development*, NY: Oxford University Press.

Hufner, K., Meyer, J. W. & Naumann, J. (1987). Comparative education policy research: A world society perspective. In Meinolf Diekes, Hans Weiler & Adriane Berthoin (eds.), *Comparative Policy Research*, Gower: WZB Publications, pp. 188-243.

Inhaber, H. (1977). Scientists and economic growth, *Social Studies of Science*, *7*:517-524.

Inkeles, A. & Smith, D. (1974). *Becoming modern: Individual change in six developing countries*, Cambridge: Harvard University Press.

Kamens, D. H. & Benavot, A. (1991). Elite knowledge for the masses: The origins and spread of mathematics and science education in national curricula, *American Journal of Education*, *99*(2):137-180.

Kamens, D. H., Meyer, J. W. & Benavot, A. (1993). The changing content of world secondary education systems, 1920-1990. Paper presented at the American Sociological Association Annual Meeting 1993, Miami, Florida.

Kelly, A. (1990). Who gets science education? In I. Varcoe, M. McNeil & S. Yearley (eds.), *Deciphering Science and Technology*, NY: MacMillan Press, pp. 50-73.

Lee, M. (1990). *Structural determinants and economic consequences of science education: A Cross-national study, 1950-1986*. Unpublished Ph.D. Dissertation, Stanford University - School of Education.

Lee, M. (1992). School science curriculum reforms in Malaysia: World influences and national context, *International Journal of Science Education*, *14*(3): 249-263.

Lee, M. & Wong, S. (1990). The provision of science education (1950-1986): Cross-national Patterns Paper presented at the Comparative and International Education Society (CIES) Annual Meeting 1990, Anaheim, California.

Lewis, K. M. (1993). Planning policy on science education in developing countries, *International Journal of Science Education*, *15*(1):1-15.

Lockheed, M. & Vespoor, A. (1990). *Improving Primary Education in Developing Countries*, Washington DC: World Bank.

Mansfield, E. (1972). Contribution of R&D to rconomic growth in the United States, *Science*, *175*:477-486.

Manue, R. (1971). Introduction, in Philip & Hadassah Gillon (eds.), *Science and Education in Developing Countries: Proceedings of the Fifth Rehovot Conference 1969*, N.Y.: Praeger Publishers.

Mayor, F. (ed.) 1982. *Scientific research and social goal: Towards a new development model*, N.Y.: Pergamon Press.

Mazrui, A. A. (1975). The African university as a multinational corporation: Problems of penetration and dependency, *Harvard Educational Review*, *45*(2): 191-210.

McClelland, D. C. (1961). *The achieving society*, NY: Free Press.

McClelland, D. C. (1969). *Motivating economic achievement*, NY: Free Press.

McEneaney, E. (1995). *Changes in the meaning of school science and mathematics: A Cross-national Analysis*. Unpublished Ph.D. Dissertation Prospectus, Stanford University - Department of Sociology.

Medvitz, A. G. (1985). *Problems in the application of science education to national development*, Nairobi, Kenya: Institute of Development Studies, University of Nairobi.

Meyer, J. W. (1987). Self and life Course: Institutionalization and its effects. In G. M. Thomas, J. W. Meyer, F. O. Ramirez & J. Boli (eds.), *Institutional Structure: Constituting State, Society, and the Individual*, Newbury Park, CA: Sage Publications, pp.242-260.

Meyer, J. W., J. Boli & G. M. Thomas. (1987). Ontology and Rationalization in the Western Cultural Account, in G. M. Thomas, J. W. Meyer, F. O. Ramirez & J. Boli (eds.), *Institutional Structure: Constituting State, Society, and the Individual*, Newbury Park, CA: Sage Publications, pp.12-40.

Meyer, J. W., M. Hannan, R. Rubinson & G. Thomas. (1979). National Economic Development, 1950-1970: Social and Political Factors, in J. Meyer & M. Hannan (eds.), *National Development and the World System*, Chicago: University of Chicago Press.

Meyer, J. W., D. H. Kamens & A. Benavot. (1992). *School knowledge for the masses: World models and nationlal primary curricular categories in the Twentieth Century*, Washington DC: The Falmer Press.

Meyer, J. W., F. O. Ramirez & Y. N. Soysal. (1992). World expansion of mass education, 1870-1970, *Sociology of Education*, 65(2): 128-149.

Midgley, M. (1992). *Science as salvation: A modern myth and its meaning*, London & NY: Routledge.
Moravcsik, M. J. (1966). Some practical suggestions for the improvement of science in developing countries, *Minerva*, *4*:381-390.
Moravcsik, M. J. (1971). Some modest proposals, *Minerva*, 9:55-65.
Moravcsik, M. J. (1987). Agenda battles and practical cooperation in science development, *Science and Technology Studies*, *5*(3/4): 122-123.
Mordi, C. (1993). Student achievement in science: A cross-cultural comparison of the Second International Science Study (SISS) Results in Nigeria, *International Journal of Science Education*, *15*(6): 685-691.
Nandy, A. (1988). Science as a reason of state. In *Science, Hegemony, and Violence: A Requiem for Modernity*, New Delhi: Oxford University Press, pp. 1-23.
Nayar, B. K. (ed.) (1976). *Science and development: Essays in various aspects of science and development, Dedicated to the Honourable Shri C. Subramanian on the Occasion of his 65 Birthday*, Bombay: Orient Longman.
Ramirez, R. O. & J. Boli. (1987). Global patterns of educational institutionalization. In Thomas, G. M., J. W. Meyer, F. O. Ramirez & J. Boli (eds.), *Institutional Structure: Constituting State, Society, and the Individual*, Newbury Park: Sage Publications, pp.150-172.
Ramirez, F. O. & G. S. Drori. (1992). The globalization of science: An institutionalist perspective. Paper presented in American Sociological Association Annual Meeting, Pittsburgh, Pennsylvania.
Ramirez, F. O. & M. Lee. 1990. Education, science, and development. Paper presented at the U.S. Department of Education Conference on the Comparative Study of Education Systems, Washington D.C.
Rosenberg, C. E. *No other gods: On science and American social thought*, Baltimore, MD: Johns Hopkins University Press.
Rutherford, J. F. Lessons from five countries. In Klein, M. S. & J. F. Rutherford (eds.), *Science Education in Global Perspective: Lessons from Five Countries*, Boulder, Colo: Westview Press, pp. 207-231.
Sagasti, F. (1973). Underdevelopment, science and technology: The point of view of the underdeveloped countries, *Science Studies*, 3:47-59.
Science Council of Canada. 1984. *Science for every student*, Report 36, Ottawa, ON: Ministry of Supply and Services.
Scott, R. W. (1987). The adolescence of institutional theory, *Administrative Science Quarterly*, *32*(3): 493-511.
Shenhav, Y. & G. S. Drori. (1988). On science, technology, and their effect on the economies of less developed countries, *Economic Quarterly*, *138*:316-323 (Hebrew.)
Shenhav, Y. & D. Kamens. (1991). The 'costs' of institutional isomorphism: Science in non-Western countries, *Social Studies of Science*, *21*:527-545.
Shrum, W. & Y. Shenhav. (1995). Science and technology in less developed countries. In S. Jassanoff, G. Markle, J. Petersen & T. Pinch (eds.), *Handbook of Science, Technology, and Society*, Newbury Park: Sage, pp. 627-651.
Thomas, G. M., J. W. Meyer, F. O. Ramirez & J. Boli (eds.) (1987). *Institutional structure: Constituting state, society, and the individual*, Newbury Park: Sage Publications.
Walters, P. B. (1989). Educational change and national economic development, *Harvard Educational Review*, *51*(1): 94-106.
Walters, P. B. & P. J. O'Connell. (1990). Post-World War II higher education, the organization of work, and changes in labor productivity in the United States, *Research in Sociology of Education and Socialization*, *9*(4): 1-23.
Walters, P. B. & R. Rubinson. (1983). Educational expansion and economic output in the United States, 1890-1969: A production function analysis, *American Sociological Review*, *48*:480-493.
World Bank. (1988). *Education in Sub-Saharan Africa: Policies for adjustment, revitalization and expansion*, Washington DC: World Bank.
Zucker, L. G. (1987). Institutional Theories of Organization, *Annual Review of Sociology*, *13*:443-464.
Zymelman, M. (1990). *Science, education, and development in Sub-Saharan Africa*, Washington D.C.: World Bank.

Prem Naidoo, Mike Savage & Kopano Taole

Chapter 4

Science Education and the Politics of Equity

It is true that we have learned that differences can be in themselves an object of respect and a source of mutual enrichment. But when they take the form of manifest inequalities, we also feel that they are injustices. This feeling, today shared by all peoples and all nations, is an undeniable sign of progress of the human conscience.
Boutros Boutros-Ghali

This chapter is composed of two parts. The authors of each part offer different perspectives on science education and equity in South Africa. In Part I, Naidoo and Savage present a critical, economic class analysis of the international pressures brought to bear on education policy and practice in Africa. In Part II, Kopano offers insight from his position within South Africa's Foundation for Research and Development (FRD), one of the governmental agencies involved with educational change in South Africa.

Part I: Analysis and Praxis

In Part I, we examine the politics of equity in the context of South Africa and Zanzibar.[57] We contend that if redress mechanisms for inequity are not directed at all aspects of society, equity will not be achieved. We also contend that as the private sector increasingly assumes responsibility for social programming, new forms of analysis and control must evolve to ensure its accountability. We suggest that science educators and science education researchers must pay more attention to the politics of equity at the macro-level (concerning matters of government) as well as the micro-levels (being politic) in any analysis of equity and redress mechanisms. To be *politic* means to be prudent and astute. Throughout Part I, we refer to government politics as politics at the macro-level, and being politic as politics at the micro-level.

Equity

Equity refers to judgement about whether or not a given state of affairs is just, that is, "principles of justice as supplementing law."[58] Arguments concerning equity are normally within a context where one social group is benefited relative to other

[57] Zanzibar is an island off the eastern coast of Africa in the Indian Ocean. The island's leading port and largest town is Zanzibar. In 1964 Zanzibar was merged with Tanganyika to form Tanzania.
[58] *The Pocket Oxford Dictionary* (1984, p. 247). Oxford: Clarendon Press.

W. W. Cobern (ed.), Socio-Cultural Perspectives on Science Education, 75–86.

groups that are usually characterised by race, ethnicity, gender, economic status, primary language and class (Harvey & Klien, 1989). The heart of equity lies in an ability to acknowledge that even though a set of actions is in accord with a set of rules, their results may be unjust (Secada, 1989).

Though related to equality, equity is a different construct. The educational literature often uses the two terms interchangeably (Grant, 1989). It is important, however, that one be clear about where equity and equality overlap as well as where they do not. Equality, according to Chambers (1985, p. 161), is "the state of being equal – the same in size, amount and value." Differences, however, are not necessarily indicators of inequity nor is sameness necessarily a sign of equity. Giving a person equality does not mean that that equity will be achieved. For example, giving all students an equal opportunity to write the same test will not achieve equity in performance. Students' performance will vary since they may not have had the same quality of instruction, studied under similar conditions, and so forth. Therefore, in order to ensure equity in performance, teachers may need to pay more attention to some students by providing differential, not equal attention. Differential access, for example, has been used as a redress mechanism to guarantee that disadvantaged groups have equitable access to programmes from which they had been previously systematically excluded (Harvey, 1989). The USA, for example, has pursued educational equity through federal policies and redress mechanisms that focus on equity regarding *access* to schooling, equity regarding the *process* of schooling and equity regarding the *outcomes* of schooling (Grant, 1989).

A concept such as equity is steeped in ideology and generates conflict (Apple, 1989). It is a political concept that can be applied to any human enterprise. Even within education, equity has many political interpretations. The concept of equity is part of a larger social context that is constantly shifting and is itself subject to ideological conflict. Thus concepts of equity continually change and context defines their meaning. It is significant that in the USA and UK equity is decreasingly linked to past group oppression and disadvantaged conditions. More and more, equity is the guarantee of individual choice under conditions of a free market that values efficient production in terms of economic, military and educational systems (Apple, 1989).

Unexamined Concerns for Equity

As important as we believe equity to be, it is not so important that it can be promoted uncritically. To illustrate this what follows is the brief record of an episode with an international donor agency determined to promote the educational well being of girls in Islamic Zanzibar.

In 1994, an international donor agency commissioned Prem Naidoo to help the ministry of education in Zanzibar to develop a proposal for a project to improve girls' participation and performance in science at the secondary school level. This can be of no surprise since of the inhabitants of Zanzibar are Moslem and the Western concern for gender equity – which is often quite at odds with traditional African and Islamic understandings of both gender and equity.

On his arrival in Zanzibar Naidoo began holding exploratory discussions about the purpose of his being there and what needed to be done. He met with the director of educational planning, a woman who he had met on a prior visit. Together they worked on the grant proposal with a group of eight senior officials including curriculum development staff, science subject inspectors, and the chief inspector of schools. The group was powerful. They were in control of education in Zanzibar. What was startling given the nature of the commission was that six of the eight officials were women! Rarely do women anywhere predominate in such positions of power, which causes one to wonder about the actual extent to which girls and women were discriminated against in Zanzibar science education. With this question in mind, Naidoo asked for data on primary and secondary schools, and specifically on science subjects. The ministry made these available the next day and in the meantime the group continued the discussion presuming that there was indeed poor participation of girls in science education at the secondary school level.

They began to brainstorm ways to overcome the problem. Interestingly, it was a man in the group who suggested affirmative action. He elaborated that policy be changed to give girls preferential access to science and suggested a quota system of 60% girls and 40% boys for science courses in secondary schools. A woman asked whether to satisfy the quota schools would accept more girls even if they performed more poorly than boys did. The man replied, "yes." His response provoked a vociferous attack from two women who objected, insisting that girls and women be treated on merit and not as tokens. They felt affirmative action would perpetuate perceptions that girls are weak at science.

The next day the census data arrived and it clearly showed that that at secondary school level *more* girls participated in science and that they *out-performed* boys. Attrition rates in science were lower for girls. Only 1 in 1000 girls dropped out for marriage and only 1 in 10 000 for pregnancy. Indeed, the data revealed that Zanzibar has a larger problem of poor participation and performance of boys in science education than of girls. When the Zanzibar team was confronted with this data they readily admitted they knew that they did not have a problem with girls' participation and performance in science. When asked why the deception, the reply was simple, "*We need the money for curriculum renewal and if we ask the donor for money to do this, we will be refused.*"

Apartheid South Africa

Apartheid became enshrined in legislation and public administration following the National Party's rise to power in 1948. It created a school system divided by race. Apartheid separated education in schools within the Republic of South Africa from those in the self-governing territories of Transkei, Ciskei, Venda and Bophutswana that are now part of the new South Africa. Education was administered through four departments: the *D*epartments of *E*ducation and *C*ulture in the House of Assembly (DEC-A) for 'whites', the House of Delegates (DEC-D) for 'Indians', the House of Representatives (DEC-R) for 'coloureds', and the *D*epartment of *E*ducation and *T*raining (DET) for 'Africans'. Some provinces had homelands within their boundaries whose education systems they administered. Schema used to classify individuals were a challenge even to those responsible for applying them in apartheid South Africa. The system used hair and other characteristics, in addition to skin pigmentation. Apartheid defined coloureds as those with mixed white and African ancestry.

The central government allocated resources to each department of education with provision to African groups being the lowest. Table 1 illustrates the per student expenditure for secondary schools in the different education departments, student-teacher ratios, teacher qualifications, and the performance of their students. With the exception of DEC-D schools, performance was clearly linked to investment in school resources. Historically, the Indian community in South Africa valued education as leading to upward economic mobility, and strongly supported their schools.

Table 1.[59] Resource Allocation and Performance in 1989 – Secondary Education

	DEC-A	DEC-D	DEC-R	DET
Per Student Expenditure in Rands	3600	2600	2100	750
Overall Student-Teacher Ratio	16	21	25	41
% Under-qualified teachers	2	4	43	87
% School Entry Cohort Passing with School-Leaving Certificate	0.85	0.84	0.30	0.14

The primary cycle of education was 7 years and the secondary cycle was 5. Children sat the School Leaving Certificate at the end of secondary school, that is grade 12. The School Leaving Certificate examined all subjects, including the sciences, at three levels namely lower, standard and higher. Students elected to sit subjects in combinations of different levels of difficulty with each level in any subject having a different examination paper. Candidates who scored well enough at the higher level gained matriculation exemption and were selected for university.

General science was compulsory for all students in their first two years of secondary school and was a combination of the biological and natural sciences. In their last three years (grades 10-12), secondary school students could choose to study physical science (chemistry and physics), biology, and mathematics though these were not compulsory subjects on the school leaving certificate. Most white, Indian and coloured schools taught science. Most African schools did not. This led to a disparity in participation rates in science subjects by the different racial groups. In 1993, 48% of white students sat physical science on the school-leaving certificate whereas only 16% of African students did so (Table 2). Comparative figures for biology were 54% of white students sat the examination and 84% of Africans (FRD, 1993). Only 25% of those African students, however, passed in biology compared with the 95% pass rate amongst white students (Table 3). A preference for biology as a soft option is a world-wide phenomenon, not only amongst South African blacks. Students generally regard the physical sciences as "tough subjects", requiring

[59] The data is from Kahn (1993, p. 8).

mathematics, more costly laboratory equipment, and leading to "non-caring professions". Also the majority of black teachers qualified to teach sciences qualified in biology. Those teachers who did not, but were forced to teach science, preferred to teach biology. For such reasons biology was a more frequently offered option in African schools.

Table 2.[60] Standard 10 Mathematics, Physical Science and Biology Enrolment Percentages by Population Group in 1993

	Biology	Mathematics	Physical science
African	84	27	16
Coloured	83	42	21
Indian	64	67	37
White	54	69	48

Table 3.[61] Standard 10 Mathematics and Science Pass Percentages by Population Group in 1993

	Biology	Mathematics	Physical Science
African	25.2	34.5	49.9
Coloured	78.0	92.7	96.5
Indian	80.0	94.8	97.7
White	95.0	98.1	98.1

Segregation also affected performance on examinations. Only 35% of African students who sat for the secondary mathematics exam in 1993 passed compared with a 98% pass rate for white students, 93% for coloured students, and 95% for Indian students. During the same year, the physical science pass rate was 50% for African students compared with a 98% pass rate for white students, 97% for coloured students, and 98% for Indian students. Approximately 1 in 310 African students who entered school in the mid 1980s obtained matriculation passes in science and mathematics and thus qualified for university entrance in science and engineering (FRD, 1993).

Thus access to higher education was unbalanced. In 1991, African students represented only 23% of enrolment in Technikons (that is technically orientated tertiary institutions). In 1991, African enrolments in universities reached 36% but most were not in faculties of science or engineering. Historically about ten times more whites than Africans completed degrees in science, despite the fact that whites represented well fewer than 10% of the population. The ratio was even more skewed

[60] Source: FRD, 1996b.

[61] Source: FRD, 1996b.

for engineers (FRD, 1993). Furthermore, racial distribution of scientists and technologists in the workplace was also unbalanced. In 1990, 4% of scientists and engineers were coloured, 5% were Indian, 9% African and 82% were white. Only 19% of scientists and engineers were women of all races; 81% were white men (FRD, 1993).

Analysis of Equity Interventions

The new South African Government has recognised the importance to development of investment in science and technology. A white paper on education and training (Department of National Education, 1994, p. 6) stressed the need to,

> raise the worker's level of general education and skill, to support the introduction of more advanced technologies, to overcome the inheritance of racial and gender stratification in the work-force, and to achieve effective worker participation in decision making and quality improvement.

A recent white paper on science and technology (Department of Arts, Culture and Science, Technology and Culture, 1996a, p. 11) argues that government,

> has a responsibility to promote science culture, science education and literacy amongst both children and adults... and influence the attainment of equity by providing incentives for disadvantage group to study mathematics, science and achieve computer literacy.

More than six recent policy papers, reviews and planning documents address problems associated with improving the quality of science and technology education, and of ensuring equal opportunities (ANC, 1993, 1994a, & 1994b; CEPD, 1993 & 1994; Department of Arts, Culture, Science and Technology, 1996a; Department of National Education, 1994). Most argue that in the past, since high quality provision was heavily polarised towards a small number of schools, the quality of most science education was poor, and that the system wasted resources. Most African students – about 80% of the school population – were denied access to the sciences since few schools taught the subjects. As a consequence, few Africans could study science at higher levels since most failed to qualify. The poor quality of science education and poor performance of African students was correctly ascribed to the legacy of apartheid policies. This inflicted a curriculum on African students that many perceived as irrelevant, difficult given the learning context, and taught by unqualified science teachers in poorly equipped schools.

The proposed interventions to redress past inequities revolve around increasing the number of science teachers in training, opening new science streams in schools that currently do not teach science, up-grading the qualifications of science teachers, developing a more relevant science curriculum, and increasing investment in facilities in African schools. Various bridging programmes, usually in tertiary institutions, have been introduced on an *ad hoc* basis. More recently these have become part of a national programme called, SYSTEM (*S*tudents and *Y*outh into *S*cience, *T*echnology, *E*ngineering and *M*athematics). Budget pressures on the amounts provincial governments can allocate to education, however, restrict these interventions to modest levels. The current priority is to equalise per capita spending on children from all communities and to provide access to those currently excluded from the compulsory basic education cycle. Therefore, a realistic scenario is that improving science and technology education is a lower priority that will have to be achieved at current cost levels. We question, however, the assumptions that form the basis of proposed interventions for achieving equity in science education and contend that additional actions will be needed.

The first assumption is that schools need, and will be allocated, more science teachers to overcome shortages and prepare for expansion of the number of African children studying science. Currently South Africa has an overall surplus of teachers as measured against target pupil teacher ratios, though there are shortages in some geographical locations and in certain subjects. However, it is not clear whether science teacher shortages as detected by science being taught by those qualified in other subjects, are real or the result of poor deployment of qualified science teachers. In any case, it is doubtful whether qualified science teachers will accept postings to other schools. Thus, newly qualified teachers likely will remain a small proportion of the total cadre of teachers until well into the next century since the transitional government currently guarantees continued employ-ment of all existing teachers. Professional development programmes will be needed to enable serving teachers to teach science effectively.

The second assumption is that increasing physical resources will have a positive effect on participation and achievement. High levels of expenditure on facilities for practical work may not result in improved performance on examinations that largely do not test skills developed by practical work. World Bank studies in other African countries demonstrate that performances on such memory-orientated examinations do not relate to laboratory facilities (Priorities and Strategies for Education, A World Bank Review, 1995). The role and nature of practical work and how it is examined must be re-thought to ensure investment in equipment is maximised.

The third assumption is that increasing participation may reduce pass rates on public examinations. The basis of this argument is an assumption that those students currently not studying science, if given an opportunity to do so, would perform on average worse than those who are enrolled. The net effect of their participation would drive down average pass rates. Indeed, it may be that those African students currently studying physical science are disproportionately drawn from those with the most persistence, enthusiasm and aptitude for the subject and that expanded access may lead to a lower overall percentage of students with these qualities. It is also possible that rapid increases in participation in science by African students may strain facilities to the point where failure rates would be comparable to, or worse than, current ones. However, we contend that we must question the content and function of examinations in the new South Africa as closely as students' performance.

The fourth assumption is that policy and re-distribution of resources are the only necessary pre-conditions for change. The new government in South Africa, driven by an ideology of non-racism, non-sexism and democracy has developed new policies to redress past inequities. However, policy is a blunt instrument that only creates an enabling environment for change – practice determines its effectiveness. We contend that curriculum content and attitudinal factors are important additional factors affecting practice. The curriculum content is in the process of being changed, but it may take decades to overcome the psychologically devastating impact of apartheid. Old habits die hard, and we suspect that changing attitudes of both teachers and students will take time. New practice will have to be grafted on existing practice that was encultured into people over a period of 40 years. Thus change will require interventions beyond dismantling apartheid policies and improving resource allocation, such as changing attitudes and behaviours as well as curriculum content, the style of its delivery in the classroom, and how it is examined.

The last assumption is that redress mechanisms aimed at improving access and inputs will bring about equity. We predict that if redress mechanisms are not directed at *all aspects* of the educational system, as well as at other aspects of society, they will not. The Zanzibar anecdote provides a case in point. The Zanzibar anecdote illustrates the politics of equity at the micro-level and raises two issues. First, improving partici-pation of marginalized groups via redress mechanisms such as affirmative action is not without contention; both from the group to be affirmed – as was the case with the senior Zanzibar women science educators – as well as from members of the group that will not be affirmed. Radicals from the affirmed group may see this as tokenism that could lead to greater oppression, while those outside the affirmed group fear that standards may drop. All stakeholders must engage in negotiations from the beginning for effective implementation of redress mechanisms.

International donors influence educational policy and practice through their financial resources. Too often that influence is exerted without adequate analysis of local issues. Though African countries may not have the same problems as other countries, international donors tend to deal with them as a homogeneous group. Gender studies and programmes are popular among donor agencies because these are Western priorities. In Africa, the more urgent issue is the financing of basic education of high quality. Thus African countries must often define their problems according to the donor agendas, as in the case of Zanzibar. However, Zanzibar officials were politically astute and made the project fit their needs by asking for a gender project that addressed general problems of curriculum renewal. More typically one finds "Teachers sidelined in new curriculum planning" which was a headline in the *Mail and Guardian* of December 24, 1996, one of South Africa's most respected newspapers. The article argues that curriculum is being developed by "outside specialists, ...mainly driven by the labour and training sector." Despite vigorous efforts to involve all stakeholders, it seems that in South Africa as well as in Zanzibar, outsiders to the school system are again defining the agenda.

Equity through Schooling

Taking up the challenge to better involve all stakeholders in the processes of education where the lack of equity is a serious problem invariably raises the concept of affirmative action. It is a commonsense type of idea for altering participation patterns that has nonetheless had a chequered history. Apartheid in South African provides an odious example of effectiveness. Afrikaners claiming centuries of English discrimination wished to redress educational and economic imbalances between white English speaking South Africans and white Afrikaans speaking South Africans. On gaining political power in 1949, Afrikaners systematically redirected government resources to the advantage of Afrikaners and the disadvantage of others (bearing in mind that the disadvantages of either Afrikaners or English South

Africans pales in comparison to the disadvantages suffered by people of colour in South Africa). Afrikaners implemented redress policies in all aspects of society over a period of 40 years. In this case, the group seeking redress represented a small proportion of the country's population and thus required relatively few resources. Apartheid redressed an imbalance, but can replacing one elite with another, or ensuring that one more small group joins the elite be deemed equitable?

In other places affirmative action regrettably has not been so effective. The USA has operated affirmative action programmes for over 30 years. In 1995, *The Los Angeles Times* reported Federal Bureau of Labor Statistics to the effect that,

> blacks increased their presence in the nation's managerial and professional ranks only meagrely from 1983 to 1993, to 6.6%...Women held 40.9% in 1983 and 47.8% a decade later. (*The Los Angeles Times*, Part A, of February 1, 1995)

It is likely that were the data further disagregated, analyses of American affirmative action programmes would show that white women and middle class African Americans have benefited most. Of all the disadvantaged groups, these are the best prepared to take advantage of affirmative action opportunities. Should the policies of the new government in South Africa similarly succeed in enabling only middle class blacks to join the current elite, could this be deemed equitable? We contend that it cannot.

Socialisation within the culture of schooling is a critical factor in school success (Apple, 1992; Ogbu, 1992) including success in school science (Cobern & Aikenhead, in press). Thus, students who enter school from social backgrounds most consistent with the culture of schooling are best able to take advantage of the opportunities made available by affirmative action programmes. It is quite likely that affirmative action programmes in the USA have best served middle class African American students and women students because most American schooling takes place in what is essentially a middle class culture. From this perspective one can see that though affirmative action provides access to schooling for many disadvantaged children, it fails to offer an equitable opportunity for school success. This failure of affirmative action is all the more trouble-some for the excuse it provides opponents to participate in yet another round of blaming the victim. Conveniently, equity becomes seen in terms of individual failure to take advantage of opportunities, not as a failure of society to prepare individuals to do so. After all, they were given the same opportunities as everybody else!

Equity must imply more than provision of equal opportunities. It must imply equal opportunities for success. This can be done by providing the disadvantaged with a more than equitable share of resources (given their history of disadvantagement). This option is not without problems. Elites in Africa (as those in the USA have done) flee from state schools when they perceive their children losing taken-for-granted. Those who perceive they are being handicapped in such ways move and Africa cannot survive such brain drains.

Another option is to change the nature of the race – not as a radical assumption as it may appear since a closer examination of the content and style of schooling in Africa, reveals that as far as quality science education is concerned, every group is being penalised. South Africa has done so. *The Electronic Mail and Guardian* (April 8, 1997) reports that:

> **The radically changed school curriculum does away with concepts like syllabus, subjects, examinations, history, geography and literature. In come eight essential "outcomes" that a learner must achieve at each stage in school, or the "learning environment" as it is now called... School pupils in South Africa can now take textbooks with them into the examination room as part of a radical change in the curriculum.**

The article continues,

> **Education will now mean developing skills... In other words, it answers two questions: 'what is the purpose of a school and why go to school?' says Emilia Potenza, a curriculum specialist... 'It is part of our revolution,' says the ruling African National Congress.**

Perhaps other African countries think that by holding onto outdated courses, they can remain part of the so-called global village. However, the reality of many Africans is that they remain very much part of their tribal village. We contend that issues of curriculum content and its style of delivery are as important factors in promoting equity and changing the nature of the race as policy change and redistribution of resources. The cultural context defines relevance and its definition changes over time. In the old South Africa, the content and style of delivery of the science education curriculum resonated with the patriarchal attitudes of apartheid. Little was to be questioned. The new South Africa requires more questioning and critical minds prepared to help solve the problems that the country is experiencing during transition. Practices in science classrooms should promote students' feeling of self-worth to help dispel feelings of inferiority that apartheid did its best to instill. The science curriculum should stem from the environment in which our students live, rather than from the conceptual structures favoured by academe. In short, we argue for a content and delivery style of science teaching that promotes democratisation at the classroom level and contend that doing so would be a sound basis for responsible citizenship as well as for future high level staff in science and technology.

A third option, that taken by the Afrikaners in South Africa, is to massively redirect resources in all aspects of society over a long period. Such a policy with reference to blacks in South Africa would be prohibitively costly, elitist and impossible to achieve as society is currently defined. Only a few blacks could ever hope to share the living standards enjoyed by most South African whites. Thus the cycle of disadvantage among the masses would continue and surely lead to despair and anger. All segments of society must participate in the debate on what constitutes equity and must have access to mechanisms for influencing the structure of society. We contend that educators must understand that since schools alone do not cause inequity, the schools alone will not solve the problems of inequity. As educators we must accept responsibility for a deeper level of analysis since equity will not be achieved through schooling alone.

Equity in Society

Though the provision of truly equal opportunities in education cannot result in equality for all, no society can afford to maintain an elitist structure. It is likely that current elites know this and will do everything they can to maintain their position. This can be seen everywhere where central, elected government is under attack. In Africa, donors press for privatization of services and cost sharing. In the long-established democracies of the UK and the USA, an important component of the platforms of political parties is weakening the role of central government.

Throughout the world, the role of capital increases as the powers of elected central government decrease. Long-term vision becomes that of capital and its intellectual training ground represented by institutions such as Oxbridge, the Ivy League schools, West Point and their oligarchic equivalents elsewhere. Tax cuts by central governments leave already poor communities dependent on the local tax base, private sector and philanthropists. As Nicholas Lemann puts it in a *Newsweek* article (April 28, 1997, p. 37) on the American president's "Summit for America's Future":

> Roughly, the moral cosmology of the moment puts voluntarism at the top, local government one rung down, then state government, and finally, cowering at the bottom, the federal government...

Some find multicultural diversity as divisive. Ethnic communities compete with each other for resources distracting from the critical analyses of entrenched class structures.[62] Furthermore, relocation by means of electronic media of the workplace to homes or to less developed parts of the world threatens to lead to fewer opportunities for labour to organise, more competition for less skilled jobs, and thus to increasing marginalization of already disadvantaged groups.

In the same issue of *Newsweek* (p 28), Jonathon Alter reports American President Clinton as saying that if the era of big government is over, the era of big citizen is just getting underway. However, we cannot assume that the role of capital in promoting equity will always be positive, witness its support to autocratic regimes in countries such as Nigeria. Throughout the world, even in industrialised countries, as social redress mechanisms increasingly become a responsibility of the private sector rather than of elected government, we witness continuing class formation of groups into those that 'have' and those that 'have not'.

With a strong role for central, elected government in promoting equity, all citizens may have a measure of control through the ballot box. Who has control when responsibility for redress mechanisms shifts to the private sector? Do shareholders decide which groups their companies should support, and how much should be given? Allan Sloan writes in the *Newsweek* article of April 28, 1997 (p. 34), "just wait until you try to get money and people out of corporate America to combat social problems." Sloan reports that in the USA between 1987 and 1995 the percentage of pretax income given for charitable activities dropped from 1.8% to 1.2% (p. 36). The article continues by quoting Colin Powell, "Yeah, I'd like to give them (corporate America) a guilt trip. You've all been talking about less government and let the market work. Well, you're the market." Through poor voter turnout, many express feelings of hopelessness to influence such trends. Disillusionment with elected government throws people back on the resources of their local, tribal communities. Monitoring the role of the private sector in promoting equity thus necessitates new forms of analysis of power structures followed by thoughtful action. New structures are needed to enable a democratic influencing and monitoring of both government and capital to ensure they play their expected social roles.

South Africa not only provides an outstanding example of effective affirmative action, namely apartheid, it also provides an example of the role of the peoples of

[62] Compare with Kopano's comments on the importance of African culture (p. 87-89) as well as the cultural issues addressed in other Chapters 7, 8, and 9.

South Africa, Europe and the USA in influencing international capital to promote equity. Coordinated social action within and without the country convinced international capital of the threat to profits of apartheid policies. International capital in turn influenced government and engaged in other actions that promoted equity such as provision of science education in black schools. In the new government, capital continues to promote equity under the Reconstruction Development Programme (RDP).

We argue that, for science educators, whether we regard ourselves as politically uninvolved, heirs of the Adam Smith tradition of liberal democracy, or as Marxists, there are implications of the growing trend to transfer social responsibilities from elected government to the private sector. As science educators, at the micro-level we can contribute to ensuring equity by assisting in development of gender, race and class inclusive curriculum materials, practices and selection procedures that promote social justice. In our classrooms we can educate young people at the macro-level to better use our science teaching to become critically reflective about social justice. Through our science teaching we can rehearse the young in ways to effectively engage in the promotion of equity. Without such analyses our actions remain fragmented, lack impact, and contribute to maintaining the *status quo* since only when politics at both the micro and macro-levels are orientated towards principles of equity will policies be developed for all social spheres including education. Only with sufficient, coordinated social pressure at micro and macro-levels will governments' politics, policies and practices impact on capital to bring social justice in observable ways.

We contend that only if principles of equity drive all spheres of a social system are such systems likely to attain equity. Education alone cannot bring about equity even if all levels of education implement appropriate redress mechanisms. The critical issue is whether social systems, that include the private sector, have the political will to allow principles of equity to drive them rather than market forces and a concept of individual rights. The challenge for science educators lies not only in classrooms. It also lies in our participation in the larger social arena for promotion of the political will and social consciousness required for achieving equity.

Kopano Taole

Part II: An Institutional Perspective

This chapter began with a statement from Boutros Boutros-Ghali, former Secretary General of the United Nations; and it is true that taking differences as an object of respect and a source of mutual enrichment is an ideal to which all peoples and all nations must strive. It is also true that this is an ideal that sadly the world has not yet realised. There is general agreement that science and technology have had a major contribution in changing the world to what it is today. By and large countries are judged by the extent to which they have made scientific and technological advances.[63] Science and technology have thus accentuated the differences between nations of the world about which Boutros Boutros-Ghali speaks, creating as it were, situations of inequalities. Mindful of this and seeking to realise the ideal of mutual enrichment, the world is challenged to create strategies through which this will be achieved. This challenge is formally taken up by the institutions of government. As noted in Part I, however, past government institutions in South Africa have deliberately worked against equity. They used science and science education to widen the socio-economic gap between blacks and whites in South Africa during the apartheid years. In Part II of this chapter post apartheid institutional policies aimed at the redress of past inequities are examined. Examples of how these inequities are dealt with are provided, followed by an analysis of the extent to which these measures are likely to succeed. Moreover, these strategies are examined against perspectives from the *social* as well as political dimensions of science and science education.

As noted in Part I there are a number of factors that ought to be accounted for when responding to questions about equity. Part I focused on *access* as a measure of equity and more will be said about access, equity, and equality. An additional factor has to do with the nature of science and worldview (Ogunniyi, 1995). One must ask whether two racial groups, possibly holding different worldviews, subjected to the same curricular intervention, can be judged in equitable circumstance. And, if there is inequity inherent in providing the same curricular intervention to two racial groups, is it fair to expect comparable achievement between the two groups? This question is likely to spark off yet another concern to which South Africa will show particular sensitivity. In a country where blacks were subjected to decades of official oppression that a culture of inferiority, it is important to interpret the worldview hypothesis with care, mindful of the politics of its use in science and science education discourse.

Apartheid: Institutional Discrimination

The history of education in South Africa, or more specifically the history of science education, does not begin with the victory of the National Party in the whites-only

[63] As noted by Adas (1989) this judgement has been at times very prejudicial.

W. W. Cobern (ed.), *Socio-Cultural Perspectives on Science Education*, 87–97.

1948 elections. This victory nonetheless forms a major landmark in the traumatic history of South Africa. The apartheid policies that were subsequently put into place are described in Innes, Kentridge & Perold (1992). The fundamental premise of white supremacy lay at the core of apartheid. This view was not held exclusively by the white minority that lived in South Africa in the 1940s. What set South African whites apart from whites elsewhere is that after they voted into office a white supremacist government, whites in South Africa kept that party in power for nearly five decades. The National Part stayed in power until finally under pressure it gave way to the present democratically elected government.

There were many ways in which the white minority government set about to entrench white supremacy. One of these was to reserve certain kinds of jobs for whites. During the apartheid years residential areas were segregated along racial lines which also led to segregated educational systems. By law there were black educational structures and it was illegal for black pupils to enroll in white schools. Apartheid created a white world in which blacks were tolerated to the extent that they provided essential services. But at the end of the day, literally, they were required by law to disappear into the slums that were created for them in the so-called townships. Apartheid thus created a black world in which a certain way of life was engendered and certain expectations were fostered.

In contrast, science education in the white educational system provided the human resources for the growing mining and manufacturing industries that formed the backbone of the country's economy. The apartheid government poured resources into providing quality laboratory equipment at both school and tertiary levels. Mining houses, one of the beneficiaries of these apartheid policies, made substantial contributions to university science and engineering. It is interesting to examine the interplay between the following factors in the context of South Africa: the need to counteract increasing world isolation; the need to militarily defend the apartheid state; the need for Afrikaner economic empowerment. World isolation drove the South African government to create its own military industry that in turn provided further opportunity for whites in science based careers.[64] The industry, structured in secrecy, was clearly out of bounds for blacks. Its financial benefits provided the economic empowerment that the Afrikaner wanted and thus science in its application to this particular industry was an instrument that created substantial economic and security imbalance between blacks and whites.

The apartheid laws affecting the military industrial economy extended to other science based industries as well. Blacks could find employment in the mining industry but law prohibited blacks from rising much beyond the level of labourer. Similar kinds of discriminatory practices were to be found in the manufacturing industries. After five decades of apartheid discrimination the cumulative effect shows in the current make up of the South African science and technology workforce. In 1992 whites made up over 70% of South Africa's workforce of 194,789 scientists and engineers in 1992.

The world of the black child stands in stark contrast to that of the white child. The black child typically goes to a school where there is no science laboratory. If

[64] The success of this industry can be measured by incidents such as the recent outcry over a proposed sale of military equipment to Syria.

there is one it is hardly used. Most black children drop science as a school subject at the first opportunity the system permits. Table 1 (page 78) shows enrolment patterns in science and mathematics during the last year of the school phase in 1993. A much higher percentage of white pupils are enrolled in these subjects than the black pupils. As a result there are also serious distortions in pass rates between black and white pupils on the school leaving public examination. These pass rates are presented in Table 2 (page 79). The patterns shown in these tables are a result of many years of institutionalised racial segregation of education in South Africa which made it possible for the government to fund education for the its goal of white domination and privilege. Christie (1992) suggests that funding disparity is at the heart of existing educational inequalities. "Whites, constituting less than 20% of the population, absorb nearly half of the state budget for education; Africans, who make up 68% of the population and 73% of school enrolment, absorb only 36% of social costs" (Christie, 1992, p. 279). Given this pattern of expenditure on education and the gap in the socio-economic condition of blacks in comparison to whites, the figures in these tables, whilst showing the enormity of the equity problem, should not come as a surprise.

The apartheid policy system that provided more resources for white pupils than it did for black pupils created unbalanced numbers of blacks and whites in science based careers. Even though a black child may have had the choice to study school science different resource levels reduced a black child's chances of succeeding. Simply put, white pupils had an environment that nurtured a science ethos. The quality of life that was attained by whites as a result of economic privilege, access to the products of science and technology in their homes, all helped to create a positive view of what science can do and brought science alive in the white pupils' lives. The pupils thus had opportunity to develop a scientific worldview – and it is important to note that this is a worldview that is grounded in Western society (Cobern, 1994). There is a strong temptation to examine more closely white pupils' acceptance of this scientific worldview and the fact that they are of European culture, but to do so would be to miss the point. For, as Cobern argues, the point is not whether this Western scientific worldview or scientific thinking is necessary; rather it is whether it must be the only way a society views the world.

White privilege and the nurturing atmosphere it provided stands in stark contrast to the circumstances that a black child had to endure. A victim of institutionalised economic disadvantage, the child typically came from a family where parents had limited education, where they were barred from certain kinds of jobs. Thus not only was there limited exposure to science in the child's life, there were also no role models in science based careers. One does not see the kind of nurturing atmosphere that the white child was exposed to. So for the black child science education happened under the following conditions: the science itself was grounded in a Western worldview, the child held a worldview that was not necessarily Western. Yet science education did little to deal with the disequilibrium that this gap in worldviews must have created in the way the child dealt with science. These conditions created a situation of imbalance in opportunity between the black child and the white child with respect to nurturing conditions for successful engagement with science.

The New Deal

In 1994 the first democratic elections resulted in a government of National Unity led by the African National Congress (ANC) as the majority party. The period since then has witnessed much policy formulation. Part of this process involved the removal of discriminatory laws that still remained on the statute books. The other was to develop a clear direction for the *new* South Africa. The new constitution of the Republic of South Africa, adopted in 1996, forms the foundation on which all policy formulation has to build. Of particular interest is the following provision relating to equity in the provision of education:

> Everyone has the right to a basic education, including adult basic education, and to further education, which the state, through reasonable measures, must make progressively available and accessible.... Everyone has the right to receive education in the official language or languages of their choice in public educational institutions where that education is reasonably practicable. In order to ensure the effective access to, and implementation of, this right, the state must consider all reasonable educational alternatives, including single medium institutions, taking into account:- equity, practicability; and the need to redress the results of past racially discriminatory laws and practices. (Republic of South Africa, 1996, p. 14)

Taking the cue from this constitutional provision, the National Department of Education has placed equity and redress high on its list of considerations that must inform policy formulation. This is captured in the following statement taken from a draft document that sets out proposals for a new curriculum framework for general and further education and training:

> The primary task of educational policy makers is the establishment of a just and equitable education and training system which is relevant, of high quality and is accessible to all learners irrespective of race, colour, gender, age, religion, ability or language. (National Department of Education, 1996, p. 5)

In similar fashion equity has emerged as an important issue in the process of transformation of higher education which is seen by government as part of the broader process of South Africa's political, social and economic transition. Accordingly, the Ministry has put forward a vision of a transformed higher education system. This system, among other things, will promote "equity of access and fair chances of success to all – irrespective of race, colour, gender, creed, age or class – seeking to realise their potential through higher education" (Ministry of education, 1997, p. 12).

Science and science education policy formulation is driven by an almost uncritical subscription to the view that science and technology hold the key the country's international economic competitiveness. Anxiety around the current state of the science and technology system has been fuelled by publications such as *The World Competitiveness Yearbook 1996*, (IMD, 1996) and the report of the *Third International Maths and science Study*. In both publications South Africa faired quite badly. This has created a growing urgency to put into place policies aimed at substantially improving the science and technology system through human resource development. The Department of Arts, Culture, science and technology (DACST, 1996b) in its white paper on science and technology makes the following point pertaining to human resource development:

> **The most pervasive effect of the system of apartheid is the legacy of inequalities generated by decades of policy interventions specifically designed to exclude the majority of South Africans from participation in social, political and economic spheres of life. Programmes need to redress the inequalities which have excluded black women and men from the mainstream of South African society.... An effective human resource development (HRD) programme in science, engineering and technology is therefore vital to redress this imbalance, to improve our economic performance and ensure the proper functioning of the National System of Innovation (NSI). Such a programme will have to address the consequences of past deliberate policies and practices that promoted racial and gender discrimination in HRD. (DACST, 1996, p. 47)**

This policy is aimed at addressing the issue of equity. There is also recognition that in order to optimise productivity and global economic, the entire population has to contribute. We thus see a human resource development policy that seeks to create opportunity for all capable South Africans to contribute to making the science and technology system function optimally for economic prosperity. This is in marked contrast to earlier policies of repression where only a small minority was active in the science and technology system. In the next section we look at examples of implementation strategies that have been employed to redress existing inequity in science and science education.

Redress Strategies

The Foundation for Research Development (FRD) is an example of an organisation that has developed and implemented a strategy to address inequity in science and technology. The FRD is one of eight science councils created by government and falls within the portfolio of the Minister of Arts, Culture, science and technology. In response to the changing social, political and economic environment, as well as the demand these changes were going to make on the science and technology system, the FRD undertook an exercise to reformulate its mission and goals. The result of this exercise is the publication *Facing the Challenge*, (FRD, 1995b) which sets out a new vision for the organisation. The mission speaks directly to the provision of human resources and expertise in science, engineering and technology. The FRD, in the way it operates, addresses human resource development through the support of research and postgraduate education in science, engineering and technology, in tertiary institutions.

Mindful of the history of racial discrimination in South Africa, the FRD has given considerable attention to activities aimed at correcting existing imbalances. Accordingly, the FRD supports activities that will, among others, achieve the following objectives:

- To contribute significantly to an increase in the number of black engineers, scientists and technologists to a level reflective of the demographic composition of society.
- To contribute to the promotion and maturation of a research culture in science, Engineering and technology at historically black universities and all technikons.
- To contribute to staff development programmes in science, Engineering and technology at historically black universities and all technikons.
- To contribute to the improvement of the quality of science education in schools and colleges of education" (FRD, 1995a, p. 7).

One strategy through which the FRD plans to achieve the above objectives is through its University Programme. The programme, which is tailor made for each Historically Black University (HBU), aims to create an environment conducive to the development and maintenance of a research culture in all HBUs. Through the programme the FRD facilitates the development of an institutional research programme with each HBU. Each programme is intended to build on identified potential that the institution has, so that the institution can develop into a centre of excellence in the identified field. In addition to support for strengthening research infrastructure in each institution, the FRD is also prepared to offer undergraduate bursaries and assistantships to promising students. The students are linked to mentors who are expected to stimulate interest in research and thereby get the students interested in postgraduate studies. Bursaries are also available for postgraduate studies under each institutional programme. The FRD also has an open Bursary and Fellowship programme. The focus of this programme is to provide support for postgraduate students with exceptional potential. However, as a corrective measure, there are bursaries that are available only to black students in the third and forth year of their undergraduate studies. Also portions of the postgraduate bursaries are designated as equity bursaries and they are available only to black students. Another measure used by the FRD to address equity redress is by including corrective action as a criterion in evaluating all research plans submitted to the organisation for funding. Under this criterion a plan is judged with respect to the extent to which black and female students will participate in the research. The plan is also judged on the extent to which the research team intends to develop linkages with HBUs.

These measures are in line with FRD's stated plan to increase substantially the number of blacks who receive support for research. Given the shrinking resource base from which research is supported, there is a stated intention that during this period of growth in black support white support will only experience minimal growth. As an example the FRD planned to support 1900 black and 1700 white students in 1996. By the year 2000, the FRD plans to increase the level of support to 3900 black and 1900 white students. These figures indicate a broader strategy of redirecting resources from one racial group to another. The strategy seems to be based on the issue of numbers: establishing a "critical mass" of black scientists (the wisdom of this strategy is addressed in the next section). A question that arises is whether this form of discrimination against whites is constitutional. Indeed, the FRD has received some complaints to the effect that it is not. This raises a further question: Whether inequities can be meaningfully addressed without some form of reverse discrimination? We shall return to this question later. We now look at the strategies that the National Department of Education has employed in dealing with past imbalances.

The department has established standardised levels of support for children in public schools. This addresses directly the funding imbalances highlighted by Christie (1992). Application of these standards has implied that some schools have had to cut staff whilst others were eligible for additional staff. The issue of staff is the easy part. A much more daunting task is the provision of infrastructure. The existing level of disparities between former white schools and black schools is very high. It is unrealistic to expect that former black school infrastructure will be improved to the current state of former white schools. The infrastructure for science teaching is even worse. This is an area where measures to address inequities can at best come on stream at a slow pace. It is to be expected that former white schools will see little or no infrastructure development whilst the department continues to address these inequities. Government seems to have chosen not to take away from the former white schools what they already have.

The department has also launched a major process of reform of the school curriculum as a form of redress. A key starting point in this reform is a commitment to outcomes-based education. Through this approach to education the country seeks to realise the vision of "a prosperous truly united, democratic and internationally competitive country with literate, creative and critical citizens, leading productive self-fulfilled lives in a country free of violence, discrimination and prejudice" (Ministry of Education, 1996a, p. 5). It is quite clear that violence, discrimination and prejudice are practices that the country will be particularly sensitive about. This is to be expected given the recent history in which they were in prevalent to unacceptably high degrees. Education is thus charged with the responsibility to contribute towards moulding a future society that will be free of these practices. The school programme under the new curriculum is organised into eight Learning Areas including the "Natural sciences." A process of consultation is currently under way and will result in a framework document that will contain outcomes for all eight learning areas. These will be further broken down into specific outcomes, out of which learning programmes will be developed.

The Centre for Education Policy Development (Perold, 1995, p. 7), "Perold Document," has described the existing science curriculum as "outmoded, academic and content-driven." The document is also concerned that the existing curriculum does not acknowledge the social influence of science and on science. Some sixteen aims for a science curriculum are listed by the Centre including *justice*, "develop the positive values and sense of responsibility needed for participation in the creation, development and maintenance of a democratic and just society..." (Perold, 1995, p. 9). There is also recognition of the multicultural make up of South African society, thus school science must provide "opportunities to recognise and explore other worldviews and systems of explanation for natural phenomena." (p. 10). The Perold Document was a first attempt by a group of educators to articulate a new science education curriculum framework for a post-apartheid South Africa. It does appear that the document has exerted influence on the formal curriculum development process taking place under the auspices of the National Department of Education. In line with the Perold document, for example, school science outcomes in the new curriculum include sensitivity to and an appreciation of the natural sciences as a value laden human enterprise. The new curriculum will also promote recognition of and acknowledgement of contributions made to the natural sciences and explanations of the natural world by different cultures, religions and societies. Speaking directly to equity, the new curriculum aims to have students identify and deal with biases and inequities implicit in and imported through the natural sciences.

There can be no doubt that the new science curriculum in South Africa aims to deal with equity issue. It remains to be seen whether this and other strategies are likely to succeed.

Will it work?

We have presented two broad strategies to address inequities in science and science education. One strategy aims to increase black participation in science and technology. The other strategy deals with the content of science education. In this section we take a closer look at each of these strategies and attempt to seek answers to the question: Will the strategy work?

With regard to black participation in science, numbers are an obvious manifestation of existing imbalances within the science and technology system of South Africa. Strategies for increasing black participation within this system therefore make sense. The human resource complement of the system is, as is to be expected, at varying skill levels. The FRD has elected to contribute to the upper end of this skill profile by supporting university researchers and granting bursaries to post-graduate students. Success of the bursary scheme depends on a number of factors. One of these is whether there are sufficient numbers of adequately prepared black students who can take up post-graduate studies. The prevailing perception is that this is not the case. This shortage of adequately prepared black students is made worse, in the view of some people, by the perception that blacks with an undergraduate degree do not struggle to get jobs. Also, a number of potential post-graduate students may be under pressure to begin to make a contribution to the economic well being of their families. In circumstances where a black potential post-graduate student might be willing to spend a few more years to get a post-graduate qualification the pressures of an immediately available job and the need to bring economic relief to the family may conspire to obstruct the FRD goals.

A second factor that could have an impact on the bursary scheme is the racial make up of the academic staff in institutions that offer post-graduate studies in science. In 1992 only a quarter of science faculty in South African universities were black (FRD, 1996). This implies that a black student enrolling for post-graduate studies in science will more likely than not have a white supervisor. It would be naïve not to acknowledge that this state of affairs impacts the potential for student success. Not to do so would be tantamount to refusing to acknowledge the impact of many decades in which perceptions were built that blacks cannot cope with the rigours of science studies. Further, it would also imply an expectation that blacks will be helped to share in what has hitherto been white privilege without some sort of resistance. The implications seem quite clear: that blacks entering into post-graduate studies in science must be mindful of the likelihood of obstacles that may be placed in their paths. Such obstacles would in turn have an impact on whether the FRD achieves its goals.

These are only two examples and it is too early to tell how seriously these impediments to redressing racial imbalances really are. There is good historical reason, however, to believe that if these factors are not successfully addressed, the FRD goals for equity will remain illusive. The other South African focus is curriculum reform for equity.

The establishment of a just and equitable education and training system is central to all efforts at transforming education and training in South Africa. It comes as no surprise that critical education has been identified as a vehicle through which this transformation will be achieved.

> If educational practice and research are to be critical, they must address conflicts and crises in society. Critical education must disclose inequalities and suppression of whatever kind. A critical education must not simply contribute to the prolonging of existing social relationships. It cannot be the means for continuing existing inequalities in society. To be critical, education must react to the critical nature of society. (Skovsmose, 1994, p. 22)

Acceptance of the perspective of critical education offers the country a powerful tool to deal with past social relationships in which blacks were marginalised. As an example from mathematics, Frankenstein (1994) suggests that a critical understanding of numerical data can prompt one to question assumptions about how society is structured, enabling one to act from a more informed position on societal structures and processes. The years of white misinformation about the black experience in South Africa, for example, can be addressed through critical education.

The newly proposed science curriculum will seek to show sensitivity to the multicultural nature of South African society. Anstey (1997) reports on how a group of grade 4 pupils when asked to name uses of water stated that it is used for healing and to get rid of evil spirits. Anstey makes the point that in the past these answers would have been unacceptable in class. But the teacher is called upon to react differently to these in terms of the new curriculum. This change in perspective speaks directly to the view that school science must be sensitive to how other cultures explain the natural world. The new curriculum allows other worldviews to find room in the school science agenda. Of course, this prompts the question of how the relationship between different worldviews in the science classroom can be envisioned. Ogunniyi (1995) seems to move from the premise that Western science remains the ultimate goal. What is at issue for him is how science education negotiates this "selectively permeable mechanism" that is a student's alternative worldview to enable him or her to learn Western science. From this perspective inequities that grow out of differences in worldview will adequately be addressed when science education has found efficient and effective ways of eliminating the disadvantage that grows out of the fact that a pupil holds a worldview other than that espoused by Western science. And, Ogunniyi adds, science education is not at that stage yet.

Ogunniyi offers one view on how science education should address equity issues relating to worldview. Cobern (1994) on the other hand, advocates for alternative constructions of science education, which are described in terms such as holistic, social, context sensitive and subjective. Whilst Cobern does not espouse an anti-science sentiment, he warns of the dangers of accepting a tight, linear science-technology-economic development model.[65] This is a warning that South Africa has good reason to heed. And, given the near hysteria surrounding the country's poor international showing in the science and technology stakes, this linear model may

[65] Also see Drori, Chapter 3 of this volume.

appear attractive. Serious doubt must, however, be raised regarding whether equity problems will be adequately addressed within such a model.

It is clear that much remains to be done to translate the vision of a just equitable society into reality through science and science education. It is also clear that in the case of creating a critical mass of black scientists the strategies employed are not taking into account factors that can compromise the success of the strategies. In the case of science education important decisions lie ahead. It is hoped that in the end the way forward will take into account growing knowledge about "anti-racism education" (Hammersley, 1995) and heed Secada's (1996) warnings about ambitious pedagogy.

References

ANC (1993). *Science and technology education and training for economic development. Proceedings of the ANC conference*. Holiday Inn Garden Court, Johannesburg, September 10-12.

ANC (1994a). *The reconstruction and development programme —a policy framework*. Johannesburg: Umanyano Publications.

ANC (1994b). *A policy framework for education and training*. Braamfontein, Johannesburg: Education Department ANC.

Anstey G. (1997, Maarch 30). Chalking up new victories. *Sunday Times*.

Apple, M. W. (1989). How equality has been redefined in the conservative restoration. In W. G. Secada (ed.), *Equity in Education*. London: Falmer Press.

Campbell P. B. (1989). Educational equity and research paradigms. In W. G. Secada (ed.), *Equity in education*. London: Falmer Press.

CEPD (1993, August). *Building the future. Overview paper for discussion: Science & technology education & training for economic development.* Produced for the ANC conference on "Science and Technology for All."

CEPD (1994). The SYSTEM initiative — Turning disadvantage into advantage, Braamfontein: author.

Christie, P. (1992). Black education. The role of the state and business. In D Innes, M Kenbridge, H. Perold (eds). *Power and profit: Politics, labour and business in South Africa*. Cape Town: Oxford University Press.

Cobern, W. W. (1994). Thinking about alternative construction of science and science education. In M. J. Glencross (ed.), *Proceedings: SAARMSE 2nd annual meeting*. Durban: SAARMSE.

Department of Arts, Culture, Science and Technology (1996a). *South Africa's Green Paper on Science and Technology - Preparing for the 21st Century*. Pretoria: author.

Department of Arts, Culture, Science and Technology (1996b). *White paper on science and technology*. Pretoria: author.

Department of National Education. (1994). *White paper 1 on education and training*. Pretoria: author.

Frankenstein M. (1994). Understanding the politics of mathematical knowledge as an integral part of becoming critically numerate. In K. Brodie, J. Strauss (eds). *Proceedingsof the AMESA 1st National Congress and Workshop*. Johannesburg: AMESA.

Foundations for Research and Development-FRD (1993). *South African science and technology indicators*. Pretoria, South Africa: author.

Foundations for Research and Development-FRD (1995a). *Announcement of new programmes*. An unpublished document. Pretoria, South Africa: author.

Foundations for Research and Development-FRD (1995b). *Facing the challenge*. An unpublished document. Pretoria, South Africa: author.

Foundations for Research and Development-FRD (1996a). *Business plan 1996 - 2000 and predetermined objectives for 1996/1997*. An unpublished document. Pretoria, South Africa: author.

Foundations for Research and Development-FRD (1996b). *South Africa science and technology indicators 1996*. South Africa: author.

Garekwe K G., Chakalisa, P. A., Taole K. (1995, May 2-5). Renegotiating the preparation of mathematics teachers: A call for a shift in paradigm. Paper presented at the *2nd National Conference on Teacher Education*. Botswana: Molepolole.

Grant C., (1989). Equity, equality, teachers and classroom life. In W. G. Secada (ed.), *Equity in education*. London: Falmer Press.

Hammersley, M. (1995) *The politics of social science research*. London: Gage Publications.
Harvey G. & Klien S. S. (1989*)*. *Understanding and measuring equity in education: A conceptual framework*. In W. G. Secada (ed.), *Equity in education*. London: Falmer Press.
Hawkins J. A. (1981). *The Oxford Dictionary*. Oxford: Clarendon Press.
IMD International (1996). *The world competitiveness yearbook-1996*. Lausanne: author.
Innes D., Kentridge, M., & Perold, H. (eds) (1992*)*. *Power and profit: Politics labour, and business in South Africa*. Cape Town: Oxford University Press.
Kahn, M., (1993). *Building the base: Report on a sector study of science education and mathematics education*. Johannesburg, South Africa: Commission of the European Communities, Pretoria and the Kagiso Trust.
Ministry of Education (1997). Draft white paper on higher education. *Government Gazette, 382*(17944).
National Department of Education (1996). *Curriculum framework for general and further education and training*. Unpublished department draft document. Pretoria: Department of Education.
Noddings, N. (1996). Equity and mathematics: Not a simple issue. *Journal of Research in Mathematics Education, 27*(5):609 - 615.
Ogunniyi, M. B. (1995). World view hypothesis and research in science education. In A. Hendricks (ed.), *Proceedings: SAARMSE Third Annual Meeting*. Cape Town: SAARMSE.
Perold, H. (ed) (1995). Curriculum frameworks for science and technology, and mathematics. Johannesburg: Centre for Education Policy Development.
Republic Of South Africa (1996). *The Constitution of the Republic of South Africa*. Cape Town: Constitutional Assembly.
Schwarz, C. M. & Seaton, M. A. (eds.), (1985). *Chambers concise usage dictionary*. London: W and R Chambers Ltd.
Secada, W. G. (1989). *Educational equity versus equality of education: An alternative conception*. In W. G. Secada (ed.), *Equity in education*. London: Falmer Press.
Secada, W. G. & Williams, B. P. (1996). Ambitious pedagogy in mathematics education: Issues for research and development. In D. Grayson (ed.). *Proceedings of the SAARMSE Fourth Annual Meeting*. Pietersburg, South Africa: SAARMSE.
Skovsmose, O. (1994). *Towards a philosophy of critical mathematics education*. Dordrecht: Kluwer Academic Press.

Kathryn Scantlebury

Chapter 5

An Untold Story: Gender, Constructivism & Science Education[66]

Philosophers of science focus on rationality and logic, not friendship and love; on prepositional knowledge and theoretical understanding, not intimate knowledge and integrative intuition. The uniqueness and complexity of individuals are viewed as problems to be overcome by science not as irreducible aspects of nature; personal feelings and relationships are taken to be impediment to objectivity, not ingredients of discovery.

J. R. Martin (1988, p. 130)

Present science education reforms, prompted by students' declining participation and achievement in science, have focused on changing the precollege science curriculum. Those changes shift the emphasis from content to process, from students' regurgitating knowledge to showing they understand the material and an attempt to show students that science is a "way of knowing" and "a process for producing knowledge" (Rutherford & Ahlgren, 1990). Science educators have utilised constructivism as a theory of knowledge to help us understand how students learn, and in particular how this theory may improve the teaching of science education. There are several different forms of constructivism: radical constructivism, trivial constructivism, social constructivism and contextual constructivism. Several researchers have outlined constructivism's limitation as a theory of knowledge (O'Loughlin, 1992; Solomon, 1994). More specifically in a discussion about gender, science education and constructivism, McComish (1995) cited five areas that were "problematic for constructivism". (1) The nature of science; (2) the purposes of education; (3) the nature of individuals; (4) how students learn; and (5) the role of teachers. She notes that these areas are inter-related and "none... can be changed in any fundamental way if corresponding changes in the others are not made" (McComish, 1995, p. 131). In this chapter I will address how students learn and the role of teachers because the rhetorical space[67] for a meaningful discussion regarding

---

[66] I gratefully acknowledge the helpful comments and suggestions from Susan Laird on earlier drafts of this chapter.

[67] Code (1995, p. ix-x) describes rhetorical spaces as "fictive but not fanciful or fixed locations, whose (tacit, rarely spoken) territorial imperatives structure and limit the kinds of utterances that can be voiced within them with a reasonable expectation of uptake and 'choral support': an expectation of being heard, understood, taken seriously."

W. W. Cobern (ed.), Socio-Cultural Perspectives on Science Education, 99–120.

gender, the nature of science and the purposes of education does not exist (Code, 1995).

Why do I suggest that there is no rhetorical space for a meaningful discussion? I will use the literature on the nature of science to provide an example of a lack of rhetorical space. Since the late 1970s critiques have proliferated regarding a masculinist bias evident not only in scientific practice, but also in the structure of scientific knowledge (Bordo, 1987; Harding, 1986, 1991; Keller & Longino, 1996; Schiebinger, 1989). Although feminists have called for the change in how science is done and an acknowledgement from scientists that their research questions and agendas are not free from bias, there has been little movement or indication that the actual practices of research science have moved in this direction. Why is this the case? One explanation is that research science is tied strongly to the economic and political discourse of Western countries and therefore to restructure science means a fundamental change in these dominant discourses (McComish, 1995). I propose that in Code's (1995) terms, there is no rhetorical space to discuss the re-figuring of science because to do so would begin to undermine fundamental economic and political needs of Western cultures (Code, 1995; McComish, 1995). Similar arguments are made for the purposes of education. Consequently, I will focus this chapter on gender issues related to constructivism on the context of areas where there does exist the potential for change, namely the understanding of how students' learn and the role of teachers.

In this chapter, I will discuss issues regarding gender and science education as related to constructivism from a social constructivist definition of the epistemology. Social constructivism accepts that learners construct their own knowledge but that knowledge is developed and influenced by their social experiences and interactions (Tobin, 1993). First, I will provide a definition of gender. Second, I will discuss how social constructivism demands attention to four areas namely: (1) knowledge making; (2) curriculum making; (3) the teaching of science and (4) the learning of science. The third section of the chapter I will review the empirical research literature as related to these four areas. In the final part of the chapter, I propose areas for future research questions science educators could pursue if we are to seriously consider gender.

Gender as a Social Construct

Before discussing gender and science education, one must clarify the term gender. Gender can be an ambiguous and misused term. Historically, gender referred only to women and was used as a different word for biological sex (Acker, 1992). To study gender was to study women. Because men were not included in the early studies about gender, it was easy to assume they had no gender. *Gender-laden* is the term used to describe these earlier studies. The distinctions of "men" and "women" disregard the variation of gender that exists and inherently ties a discussion of gender back to the different biological functions of male and female humans. Connell (1985) described this approach as categoricalism and argued that by relying on the biological differences we cannot further develop our ideas about gender. The difficulty with gender as a category is that the simple interpretation of masculine and feminine does not adequately deal with the category's variation. The wider category definitions such as Lorber (1994) describes produces a definition of gender as more of a social and relational continuum rather than categorical. In other words gender only exists as a comparative quality (a person can be either more or less masculine, or more or less feminine than someone else regardless of biological sex). For individuals, gender has been defined as a set of practices or a normative stance regarding behaviour (e.g. feminine people do this, masculine people do that (Duerst-Lahti & Kelly, 1995). As human beings we process information about others from their behaviour, and we interpret that behaviour through a variety of frameworks, including sex, race, and class (Acker, 1992).

When we begin to consider how students learn science we must take into account how they are assigned gender by themselves, peers, and teachers. Because gender is a social category it influences and impacts any social setting. Consequently, organisations and institutions, such as schools, colleges and institutions, are socially constructed entities with gender. Gendered properties of an individual or organisation can vary depending upon factors such as time and context. An example with individuals is when looking after children males may exhibit nurturing and caring behaviours. These behaviours are usually stereotyped as feminine. Although teaching is generally considered a feminized profession, within an organisation administrative practices are masculinized. Therefore, to analyse the power dynamics of a gendered setting, one must consider the social meanings of masculine and feminine and the consequences of being assigned one or the other gender (Kelly & Duerst-Lahti, 1995). As such, the social aspect of gender as a coherent set of beliefs about masculinity and femininity must be considered when analysing power differentials in sociological environments.

In educational settings, theorists have suggested different approaches to considering gender as a social construct. First, is the proposal that we should ignore gender and provide students with a gender-free educational experience. However, Houston (1985, 1996) has argued that a gender-free education reinforces the current status quo, that is, a gender-laden public education based upon male hegemony and patriachical ideology. To move beyond this situation Martin called for a *gender-sensitive education* (Martin, 1991). A gender-sensitive education is one that takes gender into account when it matters and ignores gender when it does not matter. In order to address one of the areas that McComish (1995) defined as problematic for science education, that is the nature of individuals this interpretation of gender and gender-sensitive education must further be refined to include a pluralistic view of gender, which is impacted by race/ethnicity, class, sexuality, and historical context.

Butler articulated an expanded definition for gender and noted that the concept was ambiguous even amongst feminist scholars (Butler, 1990). Because gender is a social category it not mutually exclusive of other social categories such as race, socio-economic status and/or sexuality, in discussing how gender impacts learners in science education our views of masculine and feminine need to move toward a pluralistic description of gender (Code, 1993 & 1995). Furthermore, hooks (1984) has criticised a large portion of feminist scholarship because of its focus upon white, middle-class, heterosexual individuals and institutions and while ignoring issues of race, class, and sexuality. An enhanced gender-sensitivity would accept that biological females and males may have a variety of the qualities from the traditional masculine/feminine dichotomy, and that this variation is context-specific, depending upon time, race/ethnicity and social status.

Gender/power Nexus

Interpersonal relationships are an integral part of teaching and learning. Power is an important social construct to consider in the discussion of gender, science education and constructivism because all social relations are gendered relations (Davis, 1991). Consequently, the gendered images that students and teachers hold of themselves and of others, as well as the social setting, are also influenced by power. In discussing the role of gender and the learning of science, the power distribution in a educational setting also impacts the learners' ability to construct her/his knowledge. Academics have long argued and struggled with defining power (Lipman-Blumen, 1994). Lipman-Blumen summarised the discussion by stating "most definitions of power emphasise the ability to make others conform to one's wishes, often leaving relatively ambiguous the exact origins and nature of that capacity" (Lipman-Blumen, 1984, p. 109). In accepting power, one may subordinate others and silence their voices. Delpit (1995) described five aspects that contributed to a culture of power: (1) power issues in the classroom; (2) rules of power; (3) the rules are a reflection of those who have the power; (4) knowing and understanding the power rules is important to attaining power; and (5) subordinates are more likely to recognise the people with power. The people with power are less likely to acknowledge the existence of power and their use of it than those in the subordinate group (Delpit, 1995).

Who has power in the science classroom? Firstly, teachers have power to choose the curriculum, the teaching activities and the assessment procedures in their class. Consequently, teachers can orchestrate their students' learning experiences within the classroom. However, it is possible that those learning experiences may be non or miseducative for some students (Dewey, 1963). This may be particularly true for students who do not belong to the dominant power group (that is, White males). As Rich (1979) reminds us,

> If there is any misleading concept, it is that of "co-education": that because women and men are sitting in the same classroom hearing the same lecture, reading the same books performing the same laboratories, they are receiving the same education. They are not." (Rich, 1979, p. 241).

While, teachers control the intended curriculum, the discourse of the classroom, and what is valued as learning, students can also exercise power over the learning experience. Individuals can use *macro-manipulation* and *micromanipulation* to acquire power (Lipman-Blumen, 1984). Those who have power in order to retain or increase that power employ macro-manipulation. Micromanipulation is more often used by those who are disempowered and usually occurs at the interpersonal level. Micromanipulation usually involves the gathering of intelligence using more enhanced interpretations of body language, intonation, and reading between the lines to gain more power. Teachers have macro-manipulative power over the students' grades, curricular choices and other classroom dynamics. This aspect of power is pervaded with gender dynamics and stereotypical behaviours and interpretation of behaviours.

The power dynamics in classrooms can change the educational experiences for students in non- and miseducative ways. For example in 1993, the American Association of University Women (AAUW) released a report on sexual harassment in the public schools, entitled "Hostile Hallways" (Harris, 1993). Eighty-one percent of the surveyed male and female students stated they had experienced some form of harassment. Of those incidents, the majority of the harassment occurred in hallways (66%) and classrooms (55%). The study revealed how teachers and students silently accept this behaviour.

Another example of a micromanipulation on an interpersonal level can be found in Alton-Lee, Nuthall & Patrick (1993) study of sixth-graders. Their research focused on the students' private world, especially as to how gender influences their learning. Alton-Lee et al. (1993) identified three concurrent strands influencing the teacher's intended curriculum: (1) responding to curriculum context; (2) management of the classroom culture; and (3) participating in the socio-cultural process. The intensive study of Ann, Mia, John, Joe and Ricky's private and public experiences in their 6th grade class highlighted the privileged position of males and Whites. Although Joe (a white male) did not have enough prior knowledge to make sense of the lesson, his privileged position in the classroom enabled him both publicly and privately to harass his peers without the teacher's intervention. Joe's behaviour silenced the girls, yet Ann's private answers and interactions with her friend Julia showed she had understood the teacher's intended enacted curriculum.

> The cultural climate triggered Joe's *private* (emphasis in original) racial abuse of Ricky, but allowed the boys to engage in public sexist behaviour involving sexual innuendo and verbal harassment of the girls." (Alton-Lee, et. al, 1993, p. 79).

There are less extreme examples of sexual harassment that illustrate the personal power of students. Whyte (1984) reported that girls were harassed by boys in the class to the point where they could not finish their experiments. And according to Kelly (1985), if girls are involved with boys in mixed-sex laboratory groups, they are relegated to the service roles such as equipment collector, data recorder, or laboratory cleaner. Kahle (1990) has observed that when boys have initial access to science equipment they may damage the equipment so that other students cannot complete the laboratory.

Students (especially white, male students because of their dominant position in society) are in a position to use micromanipulation and wield power in the classroom. In her study of two English high schools, Riddell (1992) observed that teachers appear to negotiate good behaviour from the boys by "by allowing them to control the physical space of classroom, the attention of the teacher and the content of lessons" (Riddell, 1992, p.146). Spender (1982) noted that when boys are not the centre of a teacher's attention they often extort attention from the teacher by refusing to behave. Thus it is no surprise that Australian teachers working with the all-male classes (in a project promoting single-sex science and mathematics experiences) requested that their professional development focus on classroom management issues (Parker & Rennie, 1995). Davidson (1996), however, described a different example of how students can use power to manipulate the classroom environment. She reported how both Latino *and* Latina high school students frustrated non-Spanish speaking teachers by refusing to speak English on certain days of the week.

The summative and differential effects of gender often determine the distribution of power in society, institutions (such as schools and universities), and in interpersonal relationships (Acker, 1992; Duerst-Lahti & Kelly, 1995). Foucault argued that power and knowledge are linked within discourse, and should be considered in a historical context (Gordon, 1980). How students construct knowledge, especially within educational settings, is influenced by numerous factors including gender, the social milieu of school and the broader societal influences on education. As science educators we need to consider how dominant masculine hegemonic triad of society, academe, and science impacts students' knowledge constructions and describe learning situations that may help all students to learn.

Science education research focuses on how students make their knowledge, the curriculum and the teaching and learning of science. Constructivism is a dominant theory in science education literature as it provides a framework for explaining how students learn science and also how teachers learn to teach. Social constructivism demands that we pay attention to knowledge making, curriculum making and the teaching and learning of science. However, empirical research illustrates how social contexts are deeply gendered. The following section describes the empirical research related to these four areas within the theoretical constructs of gender-laden, gender-free, gender-sensitive and an enhanced gender-sensitive education.

Gender-Laden Science Education

Feminist writers have noted the hegemonic male discourse and androcentricity of society (Bem, 1993; Hekman, 1990; Lorber, 1994; Spender, 1990). While the critiques of research science focused on the questions scientists ask, the interpretation of those questions, the way science is practised and the overall social context of science (Bordo, 1987; Harding, 1986, 1991; Kelly, 1987; Schiebinger, 1989). As schools are a microcosm of societal views and values it is not surprising that the current educational climate is one that is still very gender laden (Eder, Evans & Parker, 1995; Greenberg-Lake, 1991; Harris & Associates, 1993; Orenstein, 1994; Pipher, 1994; Schultz, 1991; Wellesley, 1992). Teachers' unconscious acceptance of science's gender-role stereotypes may influence their behaviours and teaching practices. Generally, teachers' behaviours and practices reinforce the concept that science is an area in which boys successfully participate more often than girls do. Unless teachers are aware of such issues, girls and boys experience a subtle gender-laden education. For example, girls receive fewer opportunities than boys to answer questions, use science equipment, and participate in science activities. Girls' work is more likely to be perceived as less important than the same work from boys (Kahle & Meece, 1994). In a patriachical society where what is masculine is defined as the 'norm', discussing how a learner constructs her/his own knowledge of science is exacerbated and compounded by the strong masculine overtones of society and the masculine image that science portrays.

Science, as taught in schools, is a process requiring logic and rational thought. To practice science successfully one must adhere to 'the scientific method' and be objective. Yet, on the other hand, the stereotypic view of femininity is that women are not rational or logical in thought, but subjective and allow emotions to rule their decisions. The stereotypic perception of 'what is science?', 'what is feminine?' and 'what is masculine?' influences students' gendered knowledge constructions and how they relate to science (Keller, 1987). Keller (1987) has argued that we cannot change the masculine image of science without also reconceptualizing masculinity and femininity. A student's gender constructions about self and subject may also impact her/his attitudes towards learning science.

What are the gendered social experiences and interactions that influence a student to learn science? What is the evidence for suggesting that school science is gender-laden? The dominant and privileged position of males (and in particular white males) in our society is reflected in our classrooms. From their birth boys and girls are treated differently based upon their sex (Bem, 1993). One consequence of girls' and boys' different experiences is that males and females enter classrooms with different knowledge, expectations and self-confidence in their ability to learn. In the 1990's the American Association of University Women released a series of reports that documented after twenty years since the inception of the Title IX legislation how girls were being 'shortchanged' in their education by our schools (Harris, 1993; Wellesley, 1992). While at the college level, the climate for women remains "chilly", with gender issues contributing to the decline of students enrolled in undergraduate science, mathematics and engineering education (Seymour & Hewitt, 1997).

When girls are publicly ignored by teachers and harassed, often sexually, by their peers how do they survive in school? What knowledge do they construct about themselves? Cohen & Blanc's study of middle school girls identified three strategies that girls use to negotiate school. The strategies are *speaking out*, *doing school*, and *crossing borders*. When girls speak out they can be labelled as troublemakers or mavericks. Adults in authority positions label a girl as a troublemaker if she is perceived as a negative leader. In contrast, her peers support a maverick. Other girls conform to society's stereotypical expectations and "do school". Some of these girls are 'schoolgirls' others as 'play schoolgirls'. Play schoolgirls pretend to conform to the school culture. Finally, there are the girls who cross borders. They may be 'schoolgirl/cool girl' or a translator. A schoolgirl/cool girl successfully negotiates different cultures (i.e., with teachers and peers within the school, as well as at home). The translator are school leaders who can make meaning and negotiate across the borders between adolescents and adults (Cohen & Blanc, 1996).

In this current climate of reform, we need to actively challenge and change the prevalent image of science as a masculine preserve, while concurrently revising the concepts of femininity and masculinity. Girls' experiences in the science classroom must change in order to enhance both their participation rates and their achievement levels. Likewise, the literature shows the deleterious effects for both boys and girls of the ways in which boys are socialised in our classrooms (Askew & Ross, 1988; Best, 1983; Salisbury & Jackson, 1995). One approach will be to ensure that science curriculum also teaches students the historical and social influences on science and challenges the cultural values.

Making of the Curriculum

Since the mid-1970's researchers have asked why there are fewer women in science and engineering than men. Berryman asked the question "Who will do science? (Berryman, 1983). Kelly (1981) called girls and women the 'missing half', while

Spender (1982) described them as 'invisible'. Since then studies commissioned by the American Association of University Women described schools with hostile hallways (Harris, 1993) and as 'shortchanging' girls on their access to education (Wellesley, 1992). In early research studies, girls were viewed as 'deficit males', the research questions asked why boys were successful in science and girls were not - what was wrong with the girls? How do we 'fix' the girls? However, others such as feminist researchers, posed a different question, they asked, "what is wrong with science?" In seeking an answer to this question, over the past two decades science education researchers have noted that students prefer to learn about how science relates to people and living things. The work of the "Girls into Science and Technology Project " (GIST) in the UK and the McClintock Collective in Australia has shown that students, especially girls, prefer science to be taught in a way that shows the relationship between people and science and the networking of those relationships with the world (Bentley & Watts, 1987; Gianello, 1989; Head, 1985; Whyte, 1986). Although studies show that male students appear to prefer physical science topics, while female students prefer biological topics, researchers have attributed these patterns to the out-of school science experiences of girls and boys (Rennie, Parker & Hutchinson, 1985). Shroyer, Backe & Powell (1995) found that middle school students enjoyed actively participating in science, although both groups described physical science topics as uninteresting. The girls and boys in this study ranked topics such as endangered species, drugs, natural events (earthquakes, volcanoes etc.) and space travel as highly interesting. While, topics with a low interest ranking were sexual reproduction, energy and natural resources and information about jobs in science. If a topic was labelled uninteresting boys attributed this to an external cause; the subject was at fault. Girls would instead say that they found the topic difficult. Teachers' can consider which science topics students' have expressed an interest in when planning their curriculum, while cognisant of the influence of different out of school educational experiences of girls' compared with boys (Rennie et al, 1983).

The diverse educational experiences for males and females continue throughout the educational pipeline. In the middle 1980's researchers began to explore the reasons why the number of science, mathematics and engineering majors begin to decline at the university level. Overall, the climate for women at the university has been described as 'chilly' (Aisenberg & Harrington, 1988; Hall, 1984; Hall & Sandler; 1984; Sandler & Shoop, 1997; Seymour & Hewitt, 1997; Sonnert, 1995; Thomas, 1990). Seymour and Hewitt (1997) described the high student attrition rates from undergraduate science, mathematics and engineering programs (SME) as "the problem iceberg'. The main reason cited for students who "switched" from SME into other fields was that the non-SME offered better education and greater intrinsic interest. Participants in Tobias' study of undergraduate science courses made similar observations regarding course content and pedagogy (Tobias, 1990). In 1989 as a response to these and other earlier studies at the college level, the Independent Colleges Offices began Project Kaleidoscope. Initially Project Kaleidoscope focused on reforming the teaching and learning of science at two-year and liberal arts colleges (Project Kaleidoscope, 1991). Five years later, this mission expanded to include four year, doctoral granting institutions. At the colleges and universities where reform efforts are successful there is 'buy-in' to the curricular and pedagogical changes by the majority of the faculty (Tobias, 1992). These efforts began from within a department rather than from outside that unit, although financial and moral support from administrators contributed to the long-term implementation of the reform efforts ultimately the faculty were the key change agents (Tobias, 1992).

The Teaching of Science

Teachers assist students in constructing their knowledge in a variety of ways. One of these is asking students questions, responding to the students' answers in order to help the student build her/his knowledge. For over a decade, researchers have shown that the students' gender and race influence the classroom interactions between teachers and students. One consequence of whole-class instruction is that there are ample opportunities for students who are risk-takers to answer a majority of the teachers' questions. Generally, boys are the risk-takers who are more likely to ask and to answer questions (Morse & Handley, 1985; Tobin & Gallagher, 1987). Tobin and Gallagher (1987) have called students who dominate the class and/or the teacher's attention *target students*. Regardless of grade level or cultural climate, target students are predominantly white males (Kahle, 1990; Kelly, 1985; Morse & Handley, 1985; Sadker & Sadker, 1994; Tobin, 1987a; Tobin & Garnett, 1987; Tobin et al., 1988; Morse & Handley, 1985; Weinekamp et. al 1987). For example, many studies of science classes report that boys were asked more and answered more questions, given more praise and time to answer questions, and received more teacher feedback than girls were. When girls interact with their teachers, the typical interaction involves discussions regarding social activities rather than content (Brophy, 1985). And more recently Roth's (1996) intensive study of a teacher's questioning patterns in her grade 4/5 classroom showed that these patterns still dominant today's classes.

How science is taught can also favour boys. Galton (1981) defines three teaching styles commonly used in science classes: problem solving, informing, and enquiring. The problem solving approach is teacher dominated with the teacher asking the students many questions often at lower-cognitive levels. In the informing style the teacher is the source of facts which he/she delivers to the students. In contrast, the third style, the enquiry method, is student-oriented. Students initiate experiments, define hypotheses, test hypotheses, and infer results. Usually girls prefer the enquiry approach, which uses more discussions and small group activities, than the other two methods use (Kahle, 1985).

In Shroyer et. al's study, the middle school students noted their preference for action-oriented activities such as labs, role playing, taking field trips, making models. They preferred to work with their friends and disliked worksheets, workbooks and textbooks. Several researchers have reported students preference for activity based science rather than textbook/workbook approaches (Lee & Burkham, 1996; Piburn & Baker, 1993; Shroyer et. al, 1995; Speering & Rennie, 1996). However, girls' interest and enjoyment of science declines during the middle school years (Speering & Rennie, 1996). While this work provides teachers and science educators an indication of the intended curriculum that would assist all students to learn science, the enacted curriculum can minimise the impact of such planning.

Alton-Lee, et al (1993) showed how students with the power to change the direction of a lesson effect their peers' knowledge construction. Because she was effectively silenced by power dynamics of the white, male privileged students Ann's teacher had no knowledge of her understanding of the topic. Ann's social interaction in developing her understanding was limited to talking quietly with her friend Julie, not the teacher, nor other peers. Mia coped with the harassment in this class by 'doing school'; she too sat quietly and completed the work, never engaging in social interactions that may have assisted her in constructing her know-ledge. As science educators we need to assist teachers in recognising that a student's gender contributes to her/his personal interactions and therefore her/his construction of knowledge and attitude toward science.

Greenfield's recent study on K-12 students' attitudes towards science, science experiences, perceptions of scientists and classroom interactions found no gender differences by grade for students' attitudes towards science. However, elementary girls liked their science curriculum better than their male peers, this attitude was reversed for the high school students (Greenfield, 1997). Congruent with many previous studies Greenfield found the familiar pattern of boys dominating classes and teachers' attention in both positive and negative ways. However, girls asked more questions than boys and teachers tended to accept girls' called-out answers more often than they did boys' answers. High school girls initiated more private contacts with their teachers than did their male peers (Greenfield, 1997). These patterns of student questioning and teacher interactions are consistent with other studies (Jones & Wheatley, 1990).[68] It appears that teachers continue to use strategies that do not engage students in science learning, and in many cases this is particularly detrimental to females who value a positive relationship with their teachers (Speering & Rennie, 1996).

If we adhere to the constructivist ideals that students create knowledge and meaning from their social interactions with their peers and teachers, then these recent studies on the socio-cultural classroom climate which show that little has changed in science classrooms are cause for concern. Science educators insist that students learn the "discourse of science' (Tobin, McRobbie & Anderson, 1997) but unless we begin to seriously examine the nature of that discourse, little will change with the enacted and implemented curriculum. Although teachers may understand that a particular teaching strategy works and believe that they are teaching that way other research suggests that this is not likely. For example, *A Splintered Vision*[69] quotes American mathematics teachers self-reporting the implementation of the NCTM teaching standards, however videotape analysis of lessons showed that teachers did not use the most appropriate teaching strategies even though they understand what they are (TIMSS, 1996).

The recent studies that show gender-stereotypes are still being inculcated in our classrooms should be no surprise. Teachers construct their knowledge about gender and science from their experiences. When one considers who takes and teaches science at tertiary-level institutions, universities are poor examples of equity in educational settings (Crozier & Menter, 1993; Scantlebury, 1994). At the university

---

[68] See Kahle & Meece (1994) for literature review.
[69] The report of the Third International Mathematics and Science Study (TIMSS).

level, science is still overwhelmingly male and teaching practices reflect the reinforcement of science as a masculine preserve (Kelly, 1985). How does a college education impact teachers' construction of knowledge? And what impact does a learner's gender have on that knowledge? The formal and hidden curriculum of a teacher's undergraduate science education will do little to change the perception that science is a masculine preserve. In most university settings, science is taught by white males in a lecture format. In the other setting for learning science, the teaching laboratories, experiments are often "cook-book" and bear no relationship or relevance to the lectures (Hewitt & Seymour, 1991; Seymour, 1997; Thomas, 1990; Tobias, 1990, 1992). In addition, women students are exposed to the very subtle and consistent nature of gender inequities, often rendered invisible to the women, their peers and the faculty by their very constancy (Hall & Sandler, 1984; Sandler & Shoop, 1997; Spender, 1982; Thomas, 1990). Thomas' (1990) study showed how female students in undergraduate physics saw themselves as the "Other" (Woolf, 1938). These students' existed on the margins of the subject because their tutors and lecturers did not know how to interact with them. The students' solution was to reject feminine values and attributes and accept the masculine ones so strongly associated with science (Thomas, 1990). This acceptance of "male as norm" is also common outside of the scientific community. Our society, and science in particular, is androcentric (Bem 1993: Byrne, 1993; Harding, 1991). Gender inequity is the "norm" and anything else is "not normal." When one first deviates from the "norm" of gender inequity, the reaction from others, particularly those favoured by the inequities, is typically swift and harsh (Sandler & Shoop, 1997). And unless gender inequity is very blatant, such as in the case of sexual harassment, it is not perceived as a problem in the classroom by either faculty or students. Women in both Thomas' (1990) and Seymour's studies (1992) experienced subtle, and not so subtle, sexual harassment from their peers, and resentment of their presence in a traditional male arena from faculty and peers (Seymour & Hewitt, 1997).

One of the consequences of this environment is that students rely upon micro-manipulation and gender stereotypes. The following example highlights how female undergraduates by manipulating the stereotype that women cannot understand physics ensured that their male professor and teaching assistant completed the laboratory assignment for them. While in a first year undergraduate physics laboratory, I focused on one lab group comprised of two young women. At the beginning of the lab they listened to the professors' explanation and instead of beginning the lab they talked between themselves about topics unrelated to the lab or the course. Five minutes later the students engaged the teaching assistant to help them. After several minutes he left, they waited for another ten minutes and called the professor over. This pattern repeated itself throughout the lab. The young women 'finished' the lab in less than two hours. When I asked the professor about them, he informed me they often were the first ones finished. These young women had learned to exploit their gender.

Gender-roles and expectations are clearly defined and students often adhere to those roles considered appropriate to their gender. In her descriptions of a social foundations course, Lewis documents the challenges that arise when implementing feminist praxis. During a discussion of male privilege and dominance, Lewis described how her male students consciously (or unconsciously) revert to male privilege while female students become concerned about isolating and ostracising their male peers (Lewis, 1990). During the course, Lewis noted that some women underwent transformations within their political stances, which had impact on their lives. A few of the men acknowledged that they had begun to 're-vise' their views. In discussing the classroom dynamics, Lewis begins to expose and to discuss the impact of the hidden curriculum. However, as Martin warns "consciousness raising is (not) any guarantee that a person will not succumb to a hidden curriculum. But still, one is in a better position to resist if one knows what is going on. Resistance to what one does not know is difficult, if not impossible (Martin, 1994, p. 167)."

The gender-blindness of teachers does not begin when they enter their own classrooms. Many teachers, both inservice and preservice, are gender-blind and believe that the equity issues are resolved. Yet, the recent reports on gender offer evidence that girls' education is inadequate compared with that of boy's (Cohen & Blanc, 1996; Eder, Evans & Parker, 1995; Greenberg-Lake, 1991; Hansen, Walker & Flom, 1995; Harris, 1993; Orenstein, 1994; Pipher, 1994; Schultz, 1991; Wellesley, 1992). Yet why should teachers take the complex issues surrounding gender equity and science seriously when science educators often ignore or trivialise these issues? The introduction to the *Professional Knowledge Standards for Science Teacher Educators* proposed by the Association for Educators of Teachers in Science (AETS) indicates that "unless prospective and practising teachers can develop the knowledge, skills and beliefs called for in the reform documents, little will change" (AETS, 1997a, p. 1). However, those standards ignored issues of equity, and thereby, the role of science educators in preparing science teachers who are sensitive to issues of equity. More recently, the organisation issued a one page position statement regarding the inclusion of undeserved populations in science education which "urges all educators to highlight integration and inclusion with regard for all students" (AETS, 1997b, p. 1). One may only infer how these issues may be highlighted from that statement. Although AAUW recommends preparing teachers who are cognisant of and able to use equitable teaching strategies, they also do not offer advice to educators as to how to prepare teachers who are sensitive to issues of gender equity (Wellesley Center for Research on Women, 1992). Recently, Veal proposed a standard for pedagogical content knowledge for the National Council for the Accreditation of Teacher Education (NCATE) that includes equity as an integral part of science teacher education (Veal, 1996). The proposed standard reads:

> The unit ensures that teacher candidates acquire an understanding of the multiple facets (student learning, assessment, and teaching methods) of pedagogical content knowledge and learn to apply these facets meaningfully in a context in which all students have an opportunity to learn (Veal, 1996, p.9).

How does a science educator ensure her/his students meet the standards? To answer that question all parties involved in the education of future science teachers may need to look in the mirror and evaluate whether the education our students receive from us is preparing them to be equitable science teachers. As Scantlebury (1995) described, this is difficult to achieve given the strong inequitable education the majority of our preservice science teachers receive in the university setting, and a challenge that Martin (1991) noted science educators avoid also.

If we are to achieve the goal of "excellence and equity" in a "science for all" educational system, science educators need to challenge teachers' constructed knowledge regarding gender, science and their interpretations of masculinity and femininity. More importantly, we also should consider the, albeit subtle and pervasive, attitudes towards gender-blindness and how those attitudes may affect our pedagogical practices, curricular choices and theoretical underpinnings of science education and the hidden curriculum in K-16 science education.

Gender-Free Education

Gender-free education is blind to the gender-laden aspect of education. Typically, philosophers have treated "the knower as a featureless abstraction" (Code, 1991). Houston has described ignoring gender as attempting to achieve a "gender-free education" (Houston, 1985). However, Code has argued that the sex of knower is fundamental in discussions of knowledge and how the learner develops viable knowledge. Teachers, administrators and parents would probably agree that a gender-free education for students is preferred over an education that emphasises traditional sex-stereotyped roles (that is, gender-laden). Gender-free education occurs when educational policies and practices are free of gender bias. This education would be one that "made active attempts to disregard gender by obliterating gender differentiations that arose within the educational sphere" (Houston, 1985, p. 359). How would an education that was considered 'gender-free' differ from the more traditional schooling experience?

Arguably, gender-free education has been the ideal that educators have striven for, and as yet, have not achieved. One reason for this may be the flaw in thinking that by treating all students in the same way we will achieve equity. Martin (1981) cautions that "the same education for both sexes will yield the same results [is based] on the assumption that sex or gender is a difference that makes no difference."

A science curriculum based upon the premise of a gender-free education would have no references to humans because in studying humans we are engage in studies related to gender. This approach would be diametrically opposite to the current science education reform. In a gender-free education, students would be called upon equally in classes, have the same access to the resources, and have the same type of interactions with their teachers. Students who show aptitude in science would be encouraged to continue their studies and all students would achieve to the maximum limit of their capabilities. Consequently, by implementing gender-free practices teachers may eliminate the present gender-laden classroom experiences of students. However, Houston (1985) notes that an education that is free from references to gender does not necessarily equate with an education that is free from gender bias.

Science education research that focuses on students' alternative conceptions shows that learners' develop strong ideas about science conceptions outside of the classroom. Singularly, students develop their concepts of masculine and feminine well before they enter school (Best, 1983; Thorne, 1993). In Best's study of young elementary children, gender-role specialisation had a large influence over their interactions with each other and adults (Best, 1983). The children accepted societal sex-roles but school did not give them an opportunity to learn how to relate to each other. Students' constructions of gender influence their social interactions with peers and adults. A gender-free science education ignores this fact.

Gender-Sensitive Education

Although many teachers believe that their classes are free from gender bias, it is difficult to provide students with a gender-free education in a subject area that has such a strong masculinist culture. Martin (1994) described a "gender-sensitive" education as an education that takes gender into account when it matters and ignores gender when it does not matter. A gender-sensitive science education would have girls and boys speaking confidently and asking academic questions. Girls would engage their teachers in intellectual conversations and take themselves seriously as scholars. All students would show respect for differences in attitudes, opinions and behaviour that could be attributed to their peers' (or teacher's) gender, race, or socio-economic status. In revising curricula, classroom dynamics, testing and assessment procedures, and policies to ensure that students receive a gender-sensitive education we may approach what Kahle (1996) described as the "ideal state" of science education.

Feminist critiques of science and undergraduate science courses have suggested teaching strategies that would make science more gender-inclusive. For example, in laboratory courses, Rosser (1990) suggested that professors could increase and expand the type of observations using both qualitative and quantitative methods in data collection. Also, they could include personal experiences of students' in discussions and pose gender as a facet of research questions. Lecture courses could encourage students to work co-operatively in groups, give essay assignments and change exams from only multiple-choice questions to a mixture of multiple-choice, short answer, and problem solving. Barad (1995) revised an advanced physics course for undergraduate majors to challenge the notion that "physicists just want to have Phun." Rosser (1995) includes case studies of curriculum and pedagogical re-visions in mathematics and the physical, computer, and environmental sciences.

At this stage, undergraduate SME reform efforts are focused on curricular reforms and changing classroom practices. In order for educators to break the hegemonic masculine discourse of university science courses, we must challenge the status quo within cultural practices of science and the reproduction of gendered stereotypes, that is, science as a masculine preserve. Yet, institutions' cultural practices are typically cyclical in nature, thereby reinforcing the status quo rather than encouraging re-conceptualisation through divergent practices (Connell, 1987).

From a constructivist view, the teacher's role is crucial for the learner and a major concern is that often teachers, students and administrators are gender-blind. However, because the sex of the learner also influences teachers' behaviours' attitudes towards the knower and expectations, teachers' gendered stereotypes need to be addressed (Lyons, 1994). Kennedy (1990) suggests that intervention programs addressing changes in teachers' classroom practices need to force teachers "to question their experiences and to question the beliefs that are based on those experiences." Previous work in gender and science education has utilised this approach. Kahle and Meece (1994) described four types of intervention programs relating girls and science education. The first type of program focused on demasculizing and demystifying science. These programs used female role models and gave students career information. The second type targeted improving girls' self-confidence and self-perceptions of their ability to do science. The third and fourth programs assisted teachers in learning instructional strategies that actively involved girls in science lessons and in understanding the skills girls need to do science. The programs that addressed teacher perceptions' and attempted to modify classroom practices appear to have been more successful in retaining girls in science than those programs that were primarily directed at the students. Recent intervention programs have gone one step further than the examples described by Kahle and Meece (1994). Not only do these programs address teaching strategies and skills, they also require teachers to reflect upon issues regarding science and gender-role stereotypes. This is achieved in several ways. First, teachers are asked to consider their perceptions of gender-role stereotypes and how these perceptions may influence their differential treatment of girls and boys in the classroom. Second, teachers review their own curricula to determine if class materials reflect the precepts of a 'gender-laden' education rather than being 'gender-sensitive'. Third, teachers are given opportunities to practice the teaching skills that previous research has shown to encourage girls to remain in science. And fourth, teachers are encouraged to become researchers in their own classrooms, collecting data and discussing gender issues. These programs can assist teachers in re-constructing their know-ledge and views regarding their views on gender and how those views may influence and impact their students' learning environment.

Enhanced Gender Sensitive Education

Knowledge making in an enhanced gender-sensitive science education.

Constructivism as a theory of knowledge suggests that learners construct their own knowledge of the world in which they live. Students' schema regarding gender are one perspective from which they interpret the world and their own role (Beall, 1993). Currently, male, White and heterosexual perspectives dominate the Western world. As Code points out credibility is an important issue when dealing with the development of ones' own knowledge. She notes how "traditionally women have access only to experience, hence not to the stuff that knowledge is made" (Code, 1990, p. 223). Within an enhanced gender-sensitive science education the role of knowledge would be more broadly defined to include not only intellect but also emotion and intuition because aspects of how people construct knowledge are "are mutually constitutive and sustaining, rather than oppositional forces in the construction of knowledge (Code, 1990, p. 47).

The role of intuitiveness in science is illustrated by Keller's biography of Barbara McClintock (Keller, 1983). McClintock discussed her emotive connections with her corn plants, and how she could understand the workings of a cell by being within the cell. Another scientist, Anna Brito identifies with her work to the point where she says 'you've got to *be* a tumor' (Goodfield, 1981, p. 226). Martin (1988) proposes that these two scientists are doing science in a different style, and a style utilising an approach that is more associated with society's feminine cultural stereotypes. In an enhanced gender-sensitive science education students' intuitive knowledge would be valued, as well as the knowledge gained from logic. The emotive, creative and intuitive aspects of learning would also be recognised as valuable knowledge for students to acquire.

The theories of post-modernism provide science educators a theoretical framework to deal with the multiple visions of truth that occur because of the multiple versions of gender that exist. In contrast to other feminist theories, post-modernism negates the long established subjection of women arguing that all theories and 'truths' are historical and context-specific. Hekman (1990) says that post-modernism's foci aids feminist thinking because it rejects the dualisms that resulted from the Enlightenment period, namely rational/irrational, subject/object, nature/culture. Also, post-modernism's view of language as "fluid and multiple" provides feminists the forum to challenge the dominant hegemonic discourse. Across the world this is not the case and we need to be cognisant of how social class, race and gender can impact and influence a student's self-concept, educational experiences and consequently their interest, aptitude and learning of science.

Curriculum making in an enhanced gender-sensitive science education

How would the science curriculum reflect and promote an enhanced gender-sensitive science education? The *Standards* provide us with the framework to articulate an ideal science curriculum and we could use that vision to develop an enhanced gender sensitive, science education curriculum for students. Rodriguez (1977), however, has criticised the *National Science Standards* for making equity invisible in the documents' text. And the initial draft of the *Professional Knowledge Standards for Science Teacher Educators* proposed by the Association for Educators of Teachers in Science (AETS, 1997) ignored equity and thereby, the role of science educators in preparing science teachers who are sensitive to issues of equity.

More recently, NSTA has published draft standards for science teacher education for teachers at the beginning, induction and professional level (NSTA, 1997). Equity is addressed within the strand on pedagogy. Beginning teachers are expected to "incorporate science teaching strategies appropriate for learners with diverse backgrounds and learning styles (NSTA, 1997). At the induction level teachers should "Plan(ing) for and regularly includes alternative activities to teach the same concept; is able to identify primary differences in learners in the student population." While a professional teacher, needs to "demonstrate a command of alternative strategies to meet diverse needs and systematically provides activities that meet those needs". The pathway these documents promote for achieving 'equity and excellence' relies more upon changing pedagogy than curricular content and context. An enhanced gender sensitive science education, however, will be difficult to achieve without a concerted effort to revise the curriculum within the context of gender. Some of the reported gender problems in adolescents' coeducational schooling may be addressed by teaching diverse topics such as foods, nutrition, the biology of eating disorders, sexuality, childbirth, child development, family psychology, hygiene, and the environment; the sorts of problems once primarily defined and investigated by teachers in home economics. Neither science nor women alone can solve such complex problems, as many home economists once seem to have hoped, but some scientific literacy is basic to understanding many of them.

The teaching and learning of science in an enhanced gender-sensitive science education.

In any discussion of the teaching and learning of science we need to recognise that the knower, teacher and knowledge produced are inseparable. What counts as knowledge in science education for the students? For the teachers? Code (1990) established a philosophical argument that illustrates that the sex of the knower cannot be ignored when considering the know-ledge that the learner constructs. Furthermore, she has challenged the assumption that knowledge is objective by suggesting that subjective knowledge is also viable knowledge. The empirical research provides a rich backdrop of information on the science topics that interest students, their preferred learning and teaching styles, as well as their opinions on the influence of socio-cultural context (Shroyer et. al, 1995). In many of these studies, students' gender has a minor impact on these broader issues. In many instances, the majority of students regardless of gender or race, prefer science topics that relate to humans taught in a manner that allows them to discuss their ideas with their colleagues

Teachers' pedagogical knowledge and pedagogical content knowledge are seen as main facets in the promoting exemplary practice. Exemplary teachers often use a variety of teaching styles but they are similar in that they have well-run classrooms, students are motivated to learn and are usually on-task (Tobin & Fraser, 1987). These teachers observed their students' behaviours and changed their teaching strategies with minimal transition time. Tobin and Fraser (1987) concluded that,

> **For most of these teachers, decisions about when to change an activity probably are intuitive and based on an accumulation of data which include experience, know-ledge of how their students learn and non-verbal cues from students. (Tobin & Fraser, 1987, p. 209).**

Using Habermas' framework (1972), Johnston & Dunne (1996) categorised the majority of research in gender and science/mathematics education as technical and practical and challenged researchers to base their research from Habermas' emancipatory perspective. They contend that this approach will begin to change the research questions that science educators pursue because the construction of gender within the research process will further inform assumption about valued knowledge, reproduction and sustainability of societal values that replicate current gender stereotypes.

Conclusion

Science education attempts to mesh two disparate areas with differing values, philosophies and goals. Possibly one reason as to why we appear so unsuccessful in changing how science is taught, what science is taught and in particular for students belonging to those groups who have traditionally not been included in science, is the lack of a rhetorical space for educators to articulate, debate and discuss science education. Within that discussion, the dualist perspectives on the characteristics of science and the alignment of these characteristics with masculinity also exasperate the role of gender and science. For the learner, whether she or he is a K-12 student, an undergraduate science major, a preservice teacher, a teacher with many years of experience, or a college level science educator, knowledge construction of gender begins at an early age and usually remains along society's strong gender-role stereotypes. This engendered knowledge can impact a teacher's behaviour in the way a teacher interacts with students, as well as the students' perceptions of teachers and their own ability to learn. Similarly for science educators, constructed views of gender influence the research questions we ask, our teaching practices and curricular choices. What questions would be posed if gender were to be taken seriously? If science educators can begin to re-define our own field to recognise the important role that gender plays in the teaching and learning of science, and the variety of meanings for any concept, we may possibly begin to dissolve the strong connections between masculinity, femininity and science — and achieve "science for all".

References

Acker, J. (1992). Gendered institutions: From sex roles to gendered institutions. *Contemporary Sociology, 21*, 565-569.

Alton-Lee, A., Nuthall, G., & Patrick, J. (1993). Reframing classroom research: A lesson from the private world of children. *Harvard Educational Review, 63*, 50-84.

Askew, S. & Ross, C. (1988). *Boys don't cry: Boys and sexism in education*. Milton Keynes: Open University Press.

Barad (1995). A feminist approach to teaching quantum physics. In S. Rosser (Ed.). *Teaching the majority. Breaking the gender barrier in science, mathematics and engineering*. (pp. 43 - 78). New York: Teachers College Press.

Beall, A. (1993). A social constructionist view of gender. In A. Beall & R Sternberg. *The psychology of gender*. (pp. 127-147). New York: Guilford Press.

Bem, S. L. (1993). *The lenses of gender: transforming the debate on sexual inequality*. New Haven: Yale University Press.

Bentley, D. & Watts, M. (1987). Courting the positive virtues: a case for feminist science In A. Kelly (Ed.) *Science for girls?* (pp. 89-99). Milton Keyes: Biddles, Ltd.

Berryman, S. E. (1983). *Who will do science?* Washington D.C.: The Rockfeller Foundation.

Best, R. (1983). *We've all got scars: what boys and girls learn in elementary school*. Bloomington: Indiana University Press.

Bordo, S. (1987). The Cartesian masculinization of thought. In S. Harding & J. O'Barr (Eds.). *Sex and scientific inquiry*. (pp. 247-264). Chicago: The University of Chicago Press.

Butler, J. (1990). *Gender trouble*. New York: Routledge.

Byrne, E. (1993). *Women and science: The snark syndrome*. Falmer Press, Washington DC.

Code, L. (1990). *What can she know? Feminist theory and the construction of knowledge*. Ithaca, NY: Cornell University Press.

Code, L. (1993). Taking subjectivity into account. In L. Alcoff & E. Potter. *Feminist epistemologies*. (pp. 15-48) . New York: Routledge.

Code, L. (1995). *Rhetorical spaces: essays on gendered locations*. New York: Routledge.

Cohen, J & Blanc, S. (1996). *Girls in the middle: working to succeed in school*. Washington DC: American Association of University Women Educational Foundation.

Connell, R. W. (1985). Theorizing gender. *Sociology, 19*, 260-272.

Connell, R. W. (1987). *Gender and power*. Stanford, CA: Stanford University Press.

Crozier, G. & Menter, I. (1993). The heart of the matter? Student teachers' experiences in school. In I. Siraj-Blatchford (Ed.), *Race, gender and the education of teachers* (pp. 94-108). Philadelphia, PA: Open University Press.

Davidson, A. (1996). *Making and molding identity in schools*. Albany: State University of New York Press.

Davis, K. (1991). Critical sociology and gender relations. In K. Davis, M. Leijenaar & J. Oldersma (Eds.) *The gender of power*. (pp. 65-86). Sage Publications.

Delpit, L. (1995). *Other people's children. Cultural conflict in the classroom*. New York: The New Press.

Duerst-Lahti, G., & Kelly, R. M. (1995). On governance, leadership and gender. In G. Duerst-Lahti & R. M. Kelly (Eds.), *Gender power, leadership and governance* (pp. 11-37). Ann Arbor: University of Michigan.

Dewey, J. (1963). *Experience and education*. New York: Collier MacMillan Publshers.

Eccles, J. (1989). Bringing young women to math and science. In M. Crawford & M. Gentry (Eds.) *Gender and thought: Psychological perspectives* (pp. 36-58). New York: Springer-Verlag.

Eder, D., Evans, C. & Parker, S. (1995). *Schooltalk: Gender and adolescent culture*. New Brunswick, NJ: Rutgers University Press.

Galton, M. (1981). Differential treatment of boy and girl pupils during science lessons. In A. Kelly (Ed.). *The Missing Half* (pp. 180-191). Manchester: Manchester University Press.

Gianello, L. (Ed.) (1988). *Getting into gear: gender-inclusive teaching strategies in science*. Canberra Curriculum Development Center.

Ginorio, A. (1995). *Warming the climate for women in academic science*. Washington DC: Association of American Colleges and Universities Program on the Status and Education of Women.

Gordon, M. (1980). (Ed.) *Power/knowledge: selected interviews and other writings, 1972-1977*. New York,: Pantheon Books.

Greenberg-Lake, The Analysis Group. (1991). *Shortchanging girls, shortchanging America*. Washington, DC: American Association of University Women.

Greenfield, T. (1996). Gender, ethnicity, science achievement and attitudes.. *Journal of research in Science Teaching, 33*, 901-934.

Greenfield, T. (1997). Gender- and grade-level differences in science interest and participation. *Science Education, 81*, 259-276.

Habermas, J. (1972). *Knowledge and human interests*. London: Heinemman.

Hall, R. & Sandler, B. (1984). *The classroom climate: A chilly one for women?* Washington, D. C: Project on the Status and Education of Women, American Association of Colleges.

Hall, R. (1984). *Out of the classroom: A chilly campus climate for women?* Washington, D. C: Project on the Status and Education of Women, American Association of Colleges.

Hansen, S., Walker, J. & Flom, B. (1995). *Growing smart: What is working for girls in school*. Washington, D. C: American Association of University Women Educational Foundation.

Harding, J. (1996). Science in a masculine straight-jacket. In L. Parker, L. Rennie, & B. Fraser (Eds.), *Gender, science and mathematics: Shortening the shadow* (pp. 129-142). Boston: Kluwer Academic Publishers Press.

Harding, S. (1986). *The science question in feminism*. Ithaca, NY: Cornell University Press.

Harding, S. (1991). *Whose science? Whose knowledge? Thinking from women's lives*. Ithaca: Cornell University Press.

Harris, L. & Associates, Inc. (1993). *Hostile Hallways: The AAUW survey on sexual harassment in America's schools*. Washington, DC: American Association of University Women.

Head, J. (1985). *The personal response to science*. Cambridge, England: Cambridge University Press.

Hekman, S. (1990). *Gender and knowledge: elements of a postmodern feminism*. Boston. Northeastern University Press.

hooks, b. (1984). *Feminist theory from margin to center*. Boston, MA. South End Press.

Houston, B. (1985). Gender freedom and the subtleties of sexist education?. *Educational Theory*, Fall, 359-369.

Houston, B. (1996). Theorizing gender: how much of it do we need? In A. Diller, B. Houston, K.P. Morgan, & A. Ayim. *The gender question in education: theory, pedagogy and politics* (pp. 75-86). Boulder: Westview Press.

Gallagher, J. & Tobin, K. (1987). Teacher management and student engagement in high school science. *Science Education, 71*(4), 535-556.

Goodfield, J. (1981). *An imagined world*. New York: Harper & Row

Johnston, J. & Dunne, M. (1996). Revealing assumptions: Problematising research on gender and mathematics and science education. In L. Parker, L. Rennie, & B. Fraser (Eds.), *Gender, science and mathematics: Shortening the shadow* (pp. 53-63). Boston: Kluwer Academic Publishers Press.

Jones, M. G. & Wheatley, J. (1990). Gender differences in teacher-student interactions in science classrooms. *Journal of Research in Science Teaching, 27*, 861-874.

Kahle, J. B. (1985). Retention of girls in science: Case studies of secondary teachers. In J. Kahle (Ed.), *Women in Science: A report form the field* (pp. 49-76). Philadelphia: Falmer Press.

Kahle, J. B. (1990). Real students take chemistry and physics: Gender issues. In K. Tobin, J. B. Kahle, & B. Fraser (Eds.), *Windows into science classrooms: Problems associated with higher-level cognitive learning* (pp. 92-134). Philadelphia, PA: Falmer Press.

Kahle, J. B. (1996). Equitable science education : A discrepancy model. In L. Parker, L. Rennie, & B. Fraser (Eds.), *Gender, science and mathematics: Shortening the shadow* (pp. 129-142). Boston: Kluwer Academic Publishers Press.

Kahle, J. B. & Meece, J. (1994). Research on gender issues in the classroom. In D. Gabel (Ed.), *Handbook of research in science teaching and learning* (pp. 542-576). Washington, DC: National Science Teachers Association.

Keller, E. F. (1983). *A feeling for the organism*. San Franscico: W.H. Freeman and Co.

Keller, E. F. (1987). Women scientist and feminist critiques of science. In S. Graubard (Ed.), *Daedalus, Learning about women: Gender, politics and power* (pp. 77-92). Cambridge, MA: American Academy of Sciences.

Keller, E. F. (1992). How gender matters: Or why it's so hard for us to count past two. In G. Kirkup & L. Smith Keller (Eds*.), Inventing women. Science, technology and gender.* (pp. 42-56). Cambridge, England: Polity Press.

Keller, E. F. & Longino, H. (1996) (Eds.). *Feminism and science*. New York: Oxford University Press.

Kelly, A. (1981). (Ed.). *The missing half: girls and science education*. Manchester: Manchester University Press.

Kelly, A. (1985). The construction of masculine science. *British Journal of Sociology of Education, 6,* 133-153.

Kelly A. (Ed.) (1987). *Science for girls?* Milton Keyes: Biddles, Ltd

Kelly, R. M., & Duerst-Lahti, G. (1995). The study of gender power and its link to governance and leadership. In G. Duerst-Lahti & R. M. Kelly (Eds.), *Gender power, leadership and governance* (pp. 11-37). Ann Arbor: University of Michigan.

Kennedy, M. M. (1990). *Policy issues in teacher education*. East Lansing: Michigan State University, National Center for Research on Teacher Education.

Lee, V. & Burkam, D. (1996). Gender differences in middle grade science achievement: subject domain, ability level and course emphasis. *Science Education, 80,* 613-650.

Lewis, M. (1990). Interrupting patriarchy: politics, resistance and transformation in the feminist classroom. *Harvard Educational Review, 60,* 467-488.

Lipman-Blumen, J. (1984). *Gender roles and power*. Englewood Cliffs, NJ: Prentice-Hall.

Lipman-Blumen, J. (1994). The existential bases of power relationships: the gender role case. In L Radtke & H. Stam (Eds.), *Power/Gender: Social relations in theory and practice* (pp. 108-135). Thousand Oaks, CA: Sage Publications.

Lorber, J. (1994). *Paradoxes of gender*. New Haven: Yale University Press.

Lyons, N. (1994). Dilemmas of knowing: ethical and epistemological dimensions of teachers' work and development. In L. Stone (Ed.), *The Education Feminism Reader* (pp. 195-220). New York: Routledge.

Lyons, L., Freitag, P., & Hewson, P. (1997). Dichotomy in thinking, dilemma in actions: Researcher and teacher perspectives on a chemistry teaching practice. *Journal of Research in Science Teaching, 27,* 861-874

Martin, J. R. (1985). *Reclaiming a conversation: The ideal of the educated woman*. New Haven: Yale University Press.

Martin, J. R. (1988). Science in a different style, *American Philosophical Quarterly, 25,* 129-140.

Martin, J. R. (1991a). The contradiction and the challenge of the educated woman. *Women's Studies Quarterly, 1,* 6-27.

Martin, J. R. (1991b). What should science educators do about the gender bias in science? In M. Matthews (Ed.), *History, philosophy, and science teaching* (pp. 151-166). Toronto: OISE Press.

Martin, J. R. (1992). *The Schoolhome: changing schools for changing families*. Boston: Harvard University Press.

Martin, J. R. (1994). *Changing the educational landscape. Philosophy, women and the curriculum*. New York, Routledge.

McComish, J. (1995). Gender issues and constructivism: Problems of theory. *Access, 13,* 129-143.

McLaren, A. & Gaskell, J. (1995). Now you see it, now you don't: Gender as an issue in school science. In J. Gaskell & J. Willinsky (Eds.), *Gender In/forms curriculum: From enrichment to transformation* (pp. 136-156). Columbia University: Teachers College Press.

Menter, I. (1989). Teaching practice stasis: racism, sexism, and school experience in initial teacher education. *British Journal of Sociology of Education, 10*, 459-473.

National Research Council. (NRC) (1990). *Fulfilling the promise: Biology education in the Nation's schools.* Washington, DC: National Academy Press.

National Research Council (NRC). (1996). *National Science Education Standards.* Washington, DC: National Academy Press.

National Science Foundation (1996). *Shaping the future: New expectations for undergraduate education in science, mathematics, engineering and technology.* Washington DC: National Science Foundation (NSF 96-139).

National Science Teachers' Association. (1997). *NSTA Standards for Science Teacher Education, Version 1.* (http://science.coe.uwf.edu/aets/draftstand.htm).

O'Loughlin, M. (1992). Rethinking science education: beyond Piagetian constructivism toward a socolcultural model of teaching and learning. *Journal of Research in Science Teaching, 29*, 791-820.

Orenstein, P. (1994). *Schoolgirls: Young women, self-esteem, and the confidence gap.* New York: Doubleday.

Parker, L. & Rennie, L. (1989). Gender issues in science education with special reference to teacher education. In *Discipline review of teacher education in mathematics and science* (pp. 230-247). Department of Education and Employment, Commonwealth of Australia, Canberra: Australian Government Publishing Service.

Parker, L., & Rennie, L. (1995). *For the sake of the girls? Final report of the Western Australian single-sex education pilot project: 1993-1994.* Perth, National Key Centre for Teaching and Research in School Science and Mathematics, Curtin University of Technology.

Piburn, M. & Baker, D. (1993). If I were the teacher...qualitative study of attitude toward science. *Science Education*; 77, 393-406.

Pipher, M. (1994). *Reviving Ophelia: saving the selves of adolescent girls.* New York: Ballantine.

Project Kaleidoscope (1991). *What works: Building natural science communities.* Washngton D.C. Stamats Communications Inc.

Rennie, L., Parker, L., & Hutchinson, P. (1985). *The effect of inservice training on teacher attitudes and primary school science classroom climates.* Research Report No. 12 Perth: University of Western Australia.

Rich, A. (1979). *On lies, secrets, and silence: selected prose, 1966-1978.* London: Norton.

Riddell, S. (1992). *Gender and the politics of the curriculum.* New York: Routledge.

Rodriguez, A. J. (1997). The dangerous discourse of invisibility: A critique of the National Research Council's "National Science Education Standards". *Journal of Research in Science Teaching, 34*, 19-37.

Rosser, S. (1990). *Female friendly science.* Elmsford, NY: Pergamon Press.

Rosser, S. (Ed.). (1995). *Teaching the majority. Breaking the gender barrier in science, mathematics and engineering.* New York: Teachers College Press.

Roth, W. M. (1996). Teacher questioning in an open-inquiry learning environment: Interactions of context, content, and student responses. *Journal of Research in Science Teaching, 33*, 709-736.

Rutherford, J. & Ahlgren, A. (1990). *Science for all Americans.* New York: Oxford University Press.

Sadker M. & Sadker, D. (1994). *Failing at fairness: How America's schools cheat girls.* New York: Scribner's.

Salisbury, J. & Jackson, D. (1995). *Challenging macho values: Practical ways of working with adolescent boys.* London: Falmer Press.

Sandler, B. & Shoop, R. (1997). (Eds.) *Sexual harassment on campus: A guide for administrators, faculty, and students.* Needham Heights: Allyn and Bacon.

Scantlebury, K. (1994). Emphasizing gender issues in the undergraduate preparation of science teachers: Practicing what we preach. *Journal of Women and Minorities in Science and Engineering, 1*, 153-164.

Scantlebury, K. (1995). Challenging gender-blindness in preservice secondary science teachers, *Journal of Science Teacher Education, 6*, 134-142.

Schiebinger, L. (1989). *The mind has no sex? Women in the origins of modern science.* Cambridge, MA: Harvard University Press.

Schultz, D. L. (1991). *Risk, resiliency, and resistance: Current research on adolescent girls.* New York: National Council for Research on Women/Ms. Foundation for Girls Initiative.

Seymour, E. (1992, February). The "problem iceberg" in science, mathematics, and engineering education: student explanations for high attrition rates. *Journal of College Science Teaching*, 230-238.

Seymour, E.& Hewitt, N. (1997). *Talking about leaving: Why undergraduates leave the sciences.* Westview Press: Boulder, CO.

Shroyer, G., Backe, K., & Powell, J. (1995). Developing a science curriculum that address the learning preferences of male and female middle level students" In D. Baker & K. Scantlebury (Eds.) *Science "coeducation": viewpoints from gender, race and ethnic perspectives.* (pp. 88-107). NARST Monograph Series, Monograph #7, Manhattan, KS: National Association of Research in Science Teaching.

Solomon, J. (1989). The social construction of school science. In R. Millar (Ed.) *Doing science: Images of science in science education.* (pp. 126-136). Philadelphia PA: Falmer Press.

Speering, W., & Rennie, L. (1996). Students' perceptions about science: the impact of transition from primary to secondary school. *Research in Science Education, 26*, 283-298.

Spender, D. (1982). *Invisible women. The schooling scandal.* London, England: Writers and Readers Publishing Cooperative Society Ltd.

Third International Mathematics and Science Study (TIMSS). (1996). *A splintered vision: An investigation of US science and mathematics education.* Lansing, MI: Michigan State University.

Thomas, K. (1990). *Gender and the subject in higher education.* London: Milton Keynes, Open University Press.

Thorne, B.(1993). *Gender play. girls and boys in school.* New Brunswick, N.J. Rutgers University Press.

Tobias, S. (1990). *They're not dumb, they're different: Stalking the second tier.* Tucson, AZ: Research Corporation.

Tobias, S. (1992). *Revitalizing undergraduate science: Why some things work and most don't.* Tucson, AZ: Research Corporation.

Tobin K. (1993) (Ed.), *The practice of constructivism in science education*. Washington, DC: American Association for the Advancement of Science Press.

Tobin, K. & Fraser, B. (Eds.). (1987). *Exemplary practice in science and mathematics education.* Perth, Western Australia, Key Centre for Teaching and Research in School Science and Mathematics (especially for women): Curtin University.

Tobin, K. & Fraser, B. (1990). What does it mean to be an exemplary science teacher? *Journal of Research in Science Teaching, 27*, 3-26.

Tobin, K. & Gallagher, J. (1987). The role of target students in the science classroom. *Journal of Research in Science Teaching, 24*, 61-75.

Tobin, K. & Garnett, P. (1987). Gender differences in science activities. *Science Education, 71*(1), 91-104.

Tobin, K., Kahle, J. B., & Fraser, B. (Eds.) (1990). *Windows into science classrooms: Problems associated with higher-level cognitive learning.* Philadelphia PA: Falmer Press.

Tobin, K., McRobbie, C. & Anderson, D. (1997). Dialectical constraints to the discursive practices of a high school physics community, *Journal of Research in Science Teaching, 34*, 491-507.

Tobin, K. & Tippins, D. (1993). Constructivism as a referent for teaching and learning. In K. Tobin (Ed.) *The practice of constructivism in science education.* (pp. 3-22). Washington, DC. American Association for the Advancement of Science Press.

Wellesley Center for Research on Women. (1992). *How schools shortchange girls: a study of major findings on girls and education.* Washington, DC: American Association of University Women.

Whyte, J. (1984). Observing sex stereotypes and interactions in the school laboratory and workshop, *Educational Review, 36*,(1), 75-86.

Whyte, J. (1986). *Girls into science and technology.* London: Routledge & Kegan Paul.

Woolf, V. (1938). *Three guineas*. New York: Harcourt Brace Jovanovich.

Marissa Rollnick

Chapter 6

The Influence of Language on the Second Language Teaching and Learning of Science

**I ascribe a basic importance to the phenomenon of language…
To speak means to be in a position to use a certain syntax,
to grasp the morphology of this or that language,
but it means above all to assume a culture,
to support the weight of a civilization.**
Frantz Fanon (1925–61)
Black Skins, White Masks

Language is a central factor to all learning. Its importance in the learning of science has often been underestimated, as there is a belief that the student's meaning will 'come through' despite language difficulties. The issue of language cannot be ignored as it impinges on the learning of science in important ways related both to attitude and cognition. In recent years it has finally come to be recognised as an important issue because of the increasing number of second language (L2) learners[70] studying science in the education systems of economically developed countries (Rosenthal 1996). As often happens, attempts to cater for this group of special needs students has also led to a realisation of the language needs of first language learners (L1) of science (e.g. Lemke, 1989).

Language in education has been an emotive and controversial issue, sometimes causing a major change in political direction in a country's future. In South Africa the 1976 Soweto uprising was sparked off by a dispute over the medium of instruction in schools. School children took to the streets to protest about the quality of their education. The 'peoples' education' movement that began with this event had a distinctive anti science undercurrent. Mathematics and science were viewed as subjects unbecoming to peoples' education as "they are divisive because not everyone can understand them" (Kahn & Rollnick, 1992).

This chapter focuses on L2 learners and will pay particular attention to the large number of these in developing countries where education continues in the tongue of former colonial powers. Although there are equivalent examples where

[70] Second language learners of science' (L2 learners) refers to learners who are studying science through a language which is not their mother tongue. In the context of this paper this will usually mean studying science through the medium of English. Conversely, 'first language learners' will refer to those studying science in their mother tongue.

W. W. Cobern (ed.), Socio-Cultural Perspectives on Science Education, 121–137.

French, Portuguese and other languages are official languages, the focus of this chapter will be entirely on learning science through English as a second language. English has become the world's most widely spoken second language. It occupies an ambivalent position in many countries where knowledge of English provides access to jobs and economic power but national pride and cultural heritage promote local languages. For science, the pressures of the economic world are clearly dominant and thus learning science is complicated by the simultaneous need to learn English.

The chapter begins by establishing a theoretical framework for language and learning. It then goes on to examine some issues in L2 science learning and then looks at some possible strategies that may be used to help students in this situation.

Theoretical Framework

To understand the challenges of learning science through a second language it is first necessary to look at theoretical perspectives of the relationship of language to thought and some fundamentals of second language acquisition theory. There are several theories about the relationship of language to thought, prominent amongst them, Piaget, Chomsky, Vygotsky and the behaviourists as exemplified by Skinner. The perspective of the behaviourist relies on a rather mechanical process of stimulus response and reinforcement, which has been convincingly challenged by Chomsky (Wilkinson, 1982, p. 209) as being simplistic. Chomsky himself believes that humans have an innate ability for acquisition of language that he calls a language acquisition device or simply, LAD (Allen & Pitlanders, 1975). This device allows for speedy acquisition of language at an early age. Chomsky believes that it would be impossible for a child to learn the underlying abstract system of language without such a device. The LAD is specific only to language and supporters of Chomsky point to the slowness of concept development in contrast to language learning to support his theory. Piaget on the other hand regards language as a tool of thought. Thought initially has its roots in actions that originate in sensory motor mechanisms and combine to make schema (Sinclair de Zwart, 1972). Language depends on these thoughts and actions and is not a separate inheritance as presumed by Chomsky.

The theorist, however, that provides the most useful explanation of the relation of language to thought is Lev Vygotsky. Although most of his work was done in the early part of this century, his ideas are still relevant today. Vygotsky (1978) regards language as a mediator of thought. As such it is an essential component of thought and the means by which we interact with our environment. In early childhood egocentric (or intra personal) speech plays a central role in problem solving. Language is in fact a simultaneous manifestation of thought. Later problems are solved silently and speech takes on more of an interpersonal role. Speech also progresses from a labelling function to an instrumental one and the importance of logical connectors emerge. Piaget believed that the disappearance of egocentric speech pointed to its disappearance from the thinking process, whereas Vygotsky believed that inner speech continues to be the mechanism by which language mediates thought.

Vygotsky contrasts visual perception, which he regards as integral, with speech or language that is of necessity sequential. Although Vygotsky developed these ideas with first language acquisition in mind, they have important consequences for L2 learners. The instrumental use of language is clearly applicable to science learning. Learners need logical connectors in their language to express their thoughts and understand the thoughts of others. Kotecha et al. (1990) quotes Strevens (1980) who points to a class of what he calls 'logico-grammatical' items, essential to the expression of complex thought relationships in English. These include words such as 'although', 'conclude' and 'contrast. Many indigenous languages either lack these connectors or express them in an entirely different way. He also characterises English as using, "*A stock of international scientific terminology based on Greek and Latin roots...*" L2 learners of science fall into two broad categories[71]:

> **Category One: those who have come to an English-speaking country having received part or all of their schooling in another language**
>
> **Category Two: those who are citizens of a multilingual country where the language of official communication and the economy is English and who are 'officially' taught at school through the medium of English.**

Category One learners are usually minorities in developed countries such as the USA. They are frequently immersed in English in the school situation and to some extent in their everyday lives. Rosenthal (1996) highlights the increasing percentage of these L2 learners In the USA. Category Two learners are usually majorities in developing countries that are former colonies. For former British colonies this language would be English. Learners encounter English for the first time at school and are expected within four or five years to learn through the medium of English. English will thus be the official language of instruction although there are signs that home language[72] instruction takes place well into secondary school. This is in spite of the fact that textbooks, written work and examinations are in English. Extracts below taken from interviews carried out with first year South African students give some indication of this. In the extracts below, "A" and "B" are students interviewed on separate occasions (Rollnick & Manyatsi, 1997).

> **Int:** **Let's go to the problems you had when you came to (University). Can you expand more about the feeling of inferiority in your social problems?**
>
> **A:** **When you hear somebody speaking English you feel I'm nothing since I can't speak like that and I have to keep alone. Our schools are poor. My English is poor. Most of the time if they feel you do not speak English well they isolate you. You come from a poor school and they come from a (formerly white) school**
>
> **B:** **I encountered problems because I didn't used to. In most of my time I didn't speak English... it seemed difficult to communicate but as we were assembled in a mixed group I tried to communicate but initially I was in a difficult problem.**

To understand the situation of both categories of L2 learners characterised above, there is a need to appreciate some fundamentals of second language acquisition theory. Two important theorists in this regard are Krashen (1982) and Cummins (1980). Rosenthal (1996) discusses these in detail.

---

[71] As stated in the introduction, this chapter focuses mostly on English as the second language.

[72] The term 'home language' is now regarded as preferable to 'mother tongue' for the language the learner speaks at home.

Krashen makes a distinction between language learning and language acquisition. The L2 learner is most often engaging in the former when trying to become competent in a second language, whereas children acquiring language for the first time are engaging in the latter. Learners usually 'monitor' their new language, a process that is more effective if time is available. So writing becomes an easier task than speaking because time is available for the monitor to operate. Speech remains more hesitant until the language acquisition process has progressed further. This distinction is clearer for category one learners. Learners in category two are often engaged in a mixture of language acquisition and language learning owing to their earlier, though ineffective start with the second language.

For category two learners the more influential issue is one raised by Cummins concerning the contrast between conversational and academic language proficiency generally thought of as academic literacy. Such learners find their greatest difficulty in expressing their ideas in writing. Experience has shown that second language learners at university make rapid progress with spoken English and with understanding spoken English, but progress with writing and reading is slower. Students can often demonstrate an oral understanding of a concept but they fail to communicate it to an examiner in a written examination. This is often evidenced by the surprise that tutors express when hearing about their students' examination performance after listening to them explaining ideas orally in tutorials (Rollnick et al., 1992).

Cummins claims that students will be fluent in everyday conversation but not in discussing academic content. Evidence such as that provided by Rollnick et al above is substantiated by academic development practice in South Africa (e.g. Moore, 1994). There is a suggestion that the main problem of these students in fact lies with writing, the reverse problem of that cited by Krashen and Rosenthal above.

A further problem elaborated by Rutherford (1993) is that of scientific language itself as a language that often transcends other language differences. The shared assumptions of scientists often make communication with a fellow scientist who speaks a different language easier than speaking to a non-scientist who speaks a common language. The problems this poses for a learner are explored later in this chapter under 'cultural differences'. Rutherford also alludes to problems with trying to express scientific ideas in African languages because of problems of vocabulary, logical connectors and multiple meanings of words. An example of the latter is the Zulu word "amandla" which translates into English as "power", "force" and "strength". There is certainly research evidence to show that these issues lead to difficulties for learners in science (Langhan, D., 1996; Nkopodi & Rutherford, 1993)

Other Factors Intertwined with language

From what has been said above, the issue of language is clearly intertwined with several other factors. A distinction has already been made between category one and category two learners above. Some writers (e.g., Kahn et al. 1992) characterise this as a difference between 'second language learners' and 'foreign language learners'. The distinction is made between an immigrant who may have had an enriched learning background and citizens in developing countries who may have suffered a disadvantaged education.

A student may be identified as having a language difficulty that could have its root in several causes and may not even be consistent across all tasks. The difficulty may be rooted in several root causes, viz.,

* **Real language difficulties caused by a second language and nothing else**
* **Conceptual difficulties**
* **Cultural differences**
* **Educational and economic disadvantage**

Real language difficulties cause by a second language and nothing else: Language may be the only difficulty experienced by category one learners. The learners in this category have no essential problem with science or academic language per se. They generally also have a home language that can be used for the teaching of science, such as Polish, German or Spanish. These languages fall into the same language group as English (Indo-European) and this may make the learning of science easier. However it is also true to say that the home and school backgrounds of these learners have also generally been ones that support the learning of science. This can be contrasted against the issues of educational and economic disadvantage, discussed below. They are thus usually in the process of converting language learning into language acquisition as defined by Krashen above. However, their success in learning science is still dependent on their success in mastering the English language. They do not have the option of abandoning the study of English should the task become too difficult.

Conceptual difficulties: There has been much research on general conceptual difficulties encountered in the learning of science. Research in the constructivist tradition draws on findings that all learners have existing ideas that may be contrary to those accepted by the scientific community and that these ideas are resistant to change (Solomon, 1983; Engel Clough et al., 1987). It has been suggested by Solomon (1984) that learners use two separate knowledge systems to achieve this. In the main, research into prior knowledge of L2 learners has shown that their alternate conceptions correspond broadly with those found among L1 learners (e.g., Hewson & Hewson, 1983, in South Africa; Ivowi, 1984 & 1986, in Nigeria; and Thijs, 1986, in Zimbabwe). This would seem consistent with Vygotsky's ideas about language as a mediator of thought. Support for this point of view comes from Stavy and Wax (1992) who found that young children's learning of biological concepts was more successful using verbal tasks than when non-verbal tasks were used.

Further support for Vygotsky's theory is provided by Inglis (1993) who worked with students bridging into tertiary education. She showed that the quality of written assignments produced by one student within a week could show vastly different language proficiency in English. She proposes that the quality of the writing is closely related the student's conceptual understanding of the content of the assignment. Thus poorly written science assignments may be evidence either of poor language proficiency or of poor conceptual understanding. This was also observed by Rollnick et al (1992), also working bridging students. Extracts from the beginning and the end of the same student's essay showed that the concepts nearer the beginning were well understood, but those at the end caused problems.

> [beginning of essay]
> The atomic theory of Democritus failed because it did not lead to quantitative predictions that could be tested, it did not have feedback from successful and unsuccessful experiments. The other reason is that people on that time were not experimental.
>
> [near the end of essay]
> Albert Einstein also discovered that electrons behaves as particle nature. He observed this when he was working on the metal emitting electrons. He saw that the electrons is emitted due to certain energy. He then concluded that electrons are particles nature.

Poor language proficiency by a student cannot thus be hastily judged as a weakness in their knowledge of the language but may be a symptom of problems due to comprehension of content.

Cultural differences: Ideas about Culture and Science Education are dealt with in other chapters of this book, but there are areas where there is a close interaction with language. Sometimes teasing the two apart is hard. The question of the definition of 'culture' has long been problematic for researchers. A useful definition is that of Rohner (1984) who rejects behaviourist definitions of culture and prefers to see it as,

> The totality of equivalent and complementary learned meanings maintained by a human population or by an identifiable segment of the population and transmitted from one generation to the next.

As in the teaching of science, definitions often create more problems than solutions and writers frequently resort to renaming the term to bring out the meaning without ambiguity. For example Hewson and Hamlyn (1983) use the term `intellectual environment' for culture.

African studies by both Western and African researchers on the influence of traditional thinking and `culture' on science learning abound (e.g., Kay, 1975; Sawyerr, 1979; Ogunniyi, 1987; Ingle & Turner, 1981). Ogawa (1986) suggests a model to explain the conflicts between traditional and scientific thinking and suggests that the process of exploring western and African traditional ideas side by side would result in a deeper understanding of both. Rollnick (1988, p. 232) sums up the contradiction with the following quotation:

> The student in Africa has one name which is used at school and another one which is used at home. There is one type of acceptable behaviour at school and one at home. There is one type of dress for school and one for home. There is a language for school and a language for home. Because of this, the student, too, becomes two people. Why not two concepts of science?

Language is often part of the learner's culture and words and their meanings have their roots in cultural background. Whorf (1956) made one of the earliest hypotheses about the relationship of language to culture. Whorf believed that forms of language are prior to, and determinant of knowledge and culture. This hypothesis in its strong form has largely been discredited, though refuting the fact that language does develop as a result of one's living context is hard. Whether culture is regarded in terms of Rohner's definition above or by the use of the term, 'intellectual environment' the element of shared understanding is clear and the means of transmission of this shared understanding is language. Language thus serves as the mediator of both thought and culture and is intimately associated with both. Making a case for language as a determinant of thought is thus difficult.

'Culture' is often cited as a source of suspicion and misconception in science but the work of Hewson and Hamlyn (1983) shows that culture can also provide a source of scientific conceptions. They found that notions of heat among North Sotho and Tswana people could be linked to the local culture and language, a clear indicator of the influence of culture on the development of prior knowledge.

A different perspective on the issue of culture is that of the 'culture of science'. Rutherford (1993) draws attention to the need to initiate L2 learners into the 'culture of science' before any meaningful learning of science can take place. She regards an understanding of the learner's indigenous culture as an important prerequisite for this to take place. L2 learners may have exposure to English as a medium of instruction in their schools but they may bring to the classroom ways of thinking which are not supportive of the study of science. One such issue is that of plagiarism. Rosenthal (1996) cites this as an example of a cultural misunderstanding in the classroom. Deckert (1993) also found that first year university science majors in Hong Kong had little familiarity with the western notion of plagiarism, poor ability to recognise it and less concern for the rights of the original writer.

Another example of this nature is cited by Rampal (1992). She examines how objective and detached scientific discourse negatively influences children's learning in oral cultures. She calls for a fundamental redefinition of scientific discourse. The issue of oral transmission of knowledge is a means of communication that goes beyond conversation. For preliterate cultures oral transmission is the means of passing knowledge from generation to generation. As members of these cultures become literate, they retain in writing many of the conventions associated with oral tradition. These conventions may be important in the way language is used to mediate thought, but not considered acceptable as scientific writing, for example personalised language may be used in texts.

Another perspective that is gaining popularity is that of Critical Language Awareness (CLA). CLA is a language pedagogy based on critical discourse theory developed by Norman Fairclough (1989) and developed in the South African context by Janks (1993). CLA examines the relationship between authority, ideology and power. It involves a critical deconstruction of text by asking such questions as "Whose texts and experiences are valued?" "Whose voices are heard?" and "Whose interests are these?" The answers to these questions frequently provide more compelling answers to educational problems than research into conceptual learning only. For example, the continuous use of male role models and constant reference to the male gender in a text may exclude female readers.

Educational and economic disadvantage: The issue of education and economic disadvantage is most starkly displayed in the South African post apartheid context. The average South African black student has emerged from a disabling education background, mostly caused by under resourced schools, poorly qualified teachers with low morale. This disabling background aggravates any language difficulties the students may have and leads to a low level of scientific knowledge (Kahn et al 1992). The existence of pre entry science courses at universities in other countries in the Southern African region suggests that problems of this nature are also to be found in those countries.

Disadvantaged backgrounds lead to a lack of academic skills overall that are often interpreted by tutors as purely language problems. The following two extracts from interviews with first year students in South Africa highlight the difficulties faced by such students (Rollnick & Manyatsi, 1997). "A" and "B" are different students interviewed on separate occasions.

A: The teachers weren't qualified for senior secondary so we worked mostly on our own. They just used to take examples from the text (sic) and put them on the board. We did our own research.

A: All experiments were demonstrated. We had one Bunsen (sic) only for the teacher's use.

Int: You say you did no experiments in school?

B: None, just theory. We were told that this happens and that happens. Even potassium permanganate, we were told about it and we looked at it at home.

Int: You mean there was none at school?

B: There was no equipment in the school. Even the biology teacher was too slow and on realising that we're behind with the syllabus, left and another one came in June.

The Medium of Instruction Debate and Indigenous Language Approaches to Education

The ambivalent status of English referred to in the introduction has made the decision about medium of instruction in schools a thorny one. On the one hand, English is regarded as indispensable for communication of science internationally and for explaining clearly the concepts of science. Many languages, it is argued, do not possess the words for scientific concepts and produce inappropriate associations when coined words are used (Isa & Maskill, 1982). Furthermore, these languages lack even the logical connectors so essential to the teaching of science. (Strevens, 1980). On the other hand it is acknowledged that expecting students to learn a new and difficult subject through the medium of a second language is unreasonable, giving them a double task of mastering both science content and language. As has been seen above, it is frequently the most disadvantaged students who are given this double task.

Much of the above argument suggests that indigenous languages are in some way deficient and that students speaking these languages 'have problems' because English is not their first language. Realising that knowledge of a second language can be an advantage in concept acquisition is important as it helps the learner to see different representations of the same ideas (Swain & Cummins, 1979; Opoku, 1983).

Different solutions have been sought to this dilemma. Many countries have opted for the colonial language. For example in West Africa, English was used in favour of the local language even from the start of primary school. (Bamgbose, 1984). Portuguese is taught from the start of school in Mozambique. In Anglophone

Southern Africa, the tendency has been to use home language instruction for the first four years and gradually switch to English medium from the fifth year onwards.

A few countries, such as Malaysia and Tanzania have opted for a local language as the medium of instruction, though Tanzania only does this in the Primary school. Some research support for indigenous language instruction comes from a large longitudinal Nigerian study (Bamgbose 1984) which found that children exposed to first language instruction performed significantly better than a control group in all subjects including English. A smaller Ghanaian study by Collison (1975) found that Ghanaian children during class discussions made far higher cognitive level statements in the home language than in English. However there is a dearth of other supporting and follow up studies, which has diminished the impact of these findings. More recently McKinley et al. (1992) worked with an indigenous language approach in a Maori community in New Zealand and found that translation was far more complex than a simple substitution of terms. Perfect meaningful translation becomes more and more difficult as the difference in culture between two societies becomes greater.

Political aspirations rather than research findings inspired policy decisions on medium of instruction such as those in Tanzania, South Africa and Malaysia. However there is much in the literature on medium of instruction, carrying with it mixed messages. Although Seddon and Waweru (1987), working with Kenyan students, found that scientific concepts were effectively transferred from one language to another, recent research by McNaught (1991) suggests that there are considerable problems around constructing meaning at the interface between Zulu and English.

Two models of instruction are in general use in US for teaching science at tertiary level (Rosenthal 1996). One is the use of English as a medium of instruction coupled with support either integrated into the science classes or as a separate ESL (English as a second language) course. The second approach is instruction through the first language while the learners master English in separate classes. As mentioned above, the latter method would not be suitable in most of the developing world due to the translation problems mentioned above. In terms of the theory of Vygotsky cited above, the second model seems preferable, but it is an expensive and difficult option to provide.

A possible compromise on medium of instruction is that of mixed language, which involves code switching (switching between two languages where the speaker has some measure of competence). Much of the work on code switching has looked into the identity and status of the listener and speaker in relation to the two languages being used Khati (1987). In more detailed studies by Poplack (1981) it was found that code switches would take place in a sentence when there was a syntactic overlap between the two languages. Poplack's work suggests that switching between languages is managed with alacrity. If this is so, then a mixed language strategy is worth investigating in the classroom situation. There is also ample evidence to suggest that this is a widespread practice in many teaching situations, despite official policy. (Watson & Bashe, 1996). The simple fact that this strategy has evolved naturally to enable L2 learners to survive suggests that it is a practical option. A basis for its use can also be argued on theoretical grounds as the home language provides a bridge or second mediator for the formulation of concepts in science.

Several studies also provide support for the effectiveness of this approach. Rollnick (1988) found that the use of a mixed language strategy enhanced the remediation of alternate conceptions of air pressure. It facilitated group communication (see below) and attempts to impose English medium only for some groups in the study met with only partial success. In a Kenyan study Cleghorn (1992) found that important ideas were conveyed more easily when the teacher did not adhere to the policy of English-only instruction. Prophet and Dow (1994) working in Botswana found that language of instruction affects concept attainment for students emerging from primary school, but not as greatly for children who are two years older. This indirectly provides evidence that home language is unofficially in use in primary school.

Medium of instruction has long been a bone of contention in the Philippines. Alvarez (1991) reports that implementation of a bilingual policy that allowed use of Filipino languages alongside English for the teaching of science. The merits of the policy are still under dispute, with several studies producing conflicting findings. However, a conclusion drawn is that the first language of children is necessary for learning science. Sutman (1992) concurs, but says that effective teaching of science to students with limited English proficiency can take place alongside the teaching of English. Learning can be enhanced by the judicious use by tutors of the learners' home language.

After years of unpopular official language policies, South Africa has finally adopted a constitution that endorses eleven official languages and passed an education act which approves multilingual classrooms. Langhan (1996) in an influential position paper provides a fresh perspective on this view. He claims that any language can be developed to take the conceptual, cognitive and linguistic load of science, mathematics and technology, providing adequate resources and official commitment are allocated to its development. He considers the use of home language or primary language appropriate if:

- the learners' primary language has not developed to the level where they have the conceptual and linguistic prerequisites for the • acquisition of literacy skills in an additional language, e.g. English;
- teachers are not well trained or competent in the use of the additional language;
- learners are not regularly exposed to the use of the additional language outside the classroom;
- there is pressure in the home or the community for literacy or language maintenance in the primary language;
- the wider community views the primary language to be of lower status than the additional language, leading to low self esteem on the part of the learners.

Clearly, the de facto situation that will emerge is a striving towards the use of English, using a multilingual approach to begin with. The success of this policy will depend very heavily on the English proficiency of the teachers in the schools but will effectively mean a mixed language approach such as that studied by Rollnick (1988) above.

An outstanding conclusion of all the above is that use of home language in the verbal traffic of the science class appears to be a natural process that aids learning.

Possible strategies to assist second language learners

Rollnick and Rutherford (1996) obtained some interesting insights taping groups of Swazi preservice teachers engaged in experimental work. Some findings on use of two languages in the classroom are provided below. SiSwati is the language spoken by Swazis. In the extracts below the SiSwati has been translated into English and represented by *italic* type. The translations are as literal as is possible without losing the English meaning. Identifiers, "A", "B", "C" and "D" all refer to students who are talking to each other.

Use of home language (SiSwati): During the discussions, code switching as characterised by Poplack (1981) often takes place. These include whole sentence switches, phrase and number switches, noun switches. For example:

A: ***What are the*** particles ***doing inside?***

C: ***It means they come out of here*** because this plastic is made of particles.

Sometimes English is used and only phrases of the type, "right", " you know" are used in home language, e.g.,

B: They are made of air. ***Isn't that so?***

With larger portions, halves of sentences were usually switched where the grammatical transition was smooth, e.g.,

B: What can you say about air particles ***when they are like this?***

Reasons for use of different language: At times, even the speaker cannot explain a change of language. At other times it is clear why a change of language has taken place. Language changes can occur when a quotation is made from the text materials. This is an obvious place for a change of language as the text materials are in English. A switch to English will often happen because a group is preparing to record something in writing. Language changes can occur when a word describing a scientific concept is needed or the English form is more compact than the SiSwati. For example:

C: Yes. ***Like in Mbabane. High atmospheric pressure goes with*** low altitude, ***right?***

B: ***Yes.***

C: ***So it means that the air would decrease in the bottle, causing it to collapse. And if you take it to a place of*** low atmospheric pressure, it will expand.

When someone repeats an explanation of something that has just been explained in English, a language change can occur, e.g.,

D: It means as he is pulling this plunger, the spaces... inside the syringe are getting loose...

C: *but they said we can also talk in SiSwati, right?*

B: *Yes.*

C: *Yeah. You see here, the time he pulled the syringe, you see the plunger is in the syringe. It settled down at the end...*

Often, use of home language can bring to the surface alternate conceptions that would otherwise not have been detected. For example:

C: *There is no space for this, so when I push it forwards, there is no air because the space is now taken up by the syringe.*

A: *And when I close here, then there is totally nothing.*

C: *When you close and I pull the syringe, since when I pull this there is supposed to be air coming in so this thing can get out since that air is not getting in, it is hard for this thing to come out then the force that is supposed to be pushing, I mean the air that is supposed to be pushing this, is not there.*

A: So the answer is...

C: *When you pull, it goes back in.*

A: There is no....

C: Pressure

A: Air pressure inside

The students believe that all air has been removed from the syringe, neglecting the small amount of air in the tip. This leads them to reason that there is a vacuum in the syringe, rather than reduced pressure.

SiSwati is also the preferred language for communicating procedural matters, such as counting collisions of marbles when operating a model to explain kinetic theory. However, the names of the numbers themselves were expressed in English.

Another way in which SiSwati is frequently used in these transcripts is by one student to clarify concepts or eliminate alternate conceptions for another in a group situation, e.g.,

C: Not necessarily that it wanted to occupy the space, the only thing is that the air pressure inside here, you see, will try to fight against this other one. The air pressure, the one of a place of a higher atmospheric pressure, this one inside is less. It is going to be defeated and it is going to be less. *Do you see, I have just explained.*

B: *What are you closing for?*

C: *You see, when we close, there is air pressure here, and there is air pressure outside. Now when you take it to a place where the pressure is the same as Manzini or Mbabane, the air pressure...* (inaudible) Which means that this thing will contract.

B: Yeah

So it can be seen that home language serves several important functions in the lesson. Its role as a mediator of thought comes out clearly in the above extracts. An important pre requisite is that the instructor can understand what is being said.

Nkopodi and Rutherford (1993) experimented with teaching materials in English that were specially mediated and designed with second language learners in mind. These worksheets still required a translation of the instructions by the teacher for their successful operation. This approach has been dubbed "translation with discretion".

This section has concentrated on the verbal aspect of science teaching. Improving the writing of second language learners calls for different strategies as will be seen in the next section.

Ways of improving writing

Rosenthal (1996) has alluded to the transition from language learning to language acquisition through monitoring. She makes the point that learners engaged in this process frequently gain proficiency quicker on writing tasks, as there is more time for the monitoring to take place. This would be the case for category one learners as defined in the early part of this chapter. However for category two learners, verbal facility comes far more easily and the battle is usually to improve academic writing, especially at the tertiary level. Three studies (Rollnick at al, 1992; Rollnick et al., 1993; Rollnick, 1994) provide examples of initiatives where the teaching of writing skills is integrated into the teaching of the science content, in this case chemistry.

The first study (Rollnick et al, 1992) describes a process approach to essay writing which includes a stage of peer review of drafts which was successful. The process approach is quite time consuming for instructors but the use of peer review helps to reduce dependence on staff time, a problem highlighted in Churms et al (1995). The challenge is to teach the learners to analyse writing critically when reviewing colleagues' work.

The second study (Rollnick et al 1993) was a development from the essay writing exercise described above. Groups of students were required to produce posters on a topic of chemistry and make an oral presentation. So this was an exercise both in writing and oral presentation. This was thought to be a better preparation for the type of writing called for in examinations. The type of condensation of information required for presentation on a poster requires a far higher level of comprehension than when space is less limited for presentation. This was displayed by the performance of the students on the examination question that was set to test this task.

The third study (Rollnick, 1994) found that the use of E-mail in communicating chemistry content motivates students to express their chemistry ideas in writing. This medium also provided a good means or promoting metacognitive awareness. The effect of the e-mail milieu on student writing was of great interest. The language of the messages suggested that they were writing with readers in mind. This is generally regarded as a major step forward in teaching the novice writer to express himself or herself. It was also obvious from both their style and way in which they exposed their ideas that they felt themselves to be in a non-threatening situation. It is possible that the use of this medium approximates more closely verbal interaction. This provides an 'oral' means of communicating science content as evidenced by the differential use of register in the extract below. Note the change from the familiarity of the greeting and first statement to the more formal language used to talk about the science:

> **Hello Doc!**
>
> **Before I go any further, I would like to make it quite clear that I don't think what will follow is a good answer to your question. A lot of thinking hasn't helped much.**
>
> **According to the 2nd law of thermodynamics, the entropy change in the universe is greater than zero. This means that the sum of entropy changes of systems and their surroundings is positive. Therefore naturally, systems tend to a state of disorder (i.e. higher entropy). We have also learnt that when a system approaches equilibrium, its entropy increases. So a state of equilibrium is a state of high disorder.**

The main message is that writing be for a purpose. The learner must have a need to communicate content to readers. This point has also been highlighted by Raimes (1991) who says that writing should be for a purpose and not a mere exercise. This message is connected to overall communication and links to the teaching of all language skills in science.

Teaching of other language skills in Science

The skills, which can be loosely grouped under this heading, include writing as discussed above. They also include a range of skills that have been previously taken for granted in a university setting, but have not necessarily been taught explicitly. Kotecha et al (1990) list these as follows:

> **reading in the sciences, note taking and listening, summarising skills, information gathering, relating language to work and real life models, differences between everyday language and scientific language, interpretation of exam questions and examination technique, report writing, including the use of the passive tense, graphical presentation of information, logical expression of ideas, description and argument, presenting findings orally, sometimes to video, integrated problem solving tasks, constructing and expressing points of view in role play.**

These skills can be taught in a separate course but are more successfully acquired in context, integrated with the science content (Rollnick, 1991). Many of these integrated exercises first made their appearance in contextualised courses for first language learners in economically developed countries. They were described as active learning strategies and included activities such as role plays, creative writing, information searches, wall displays, comprehension exercises called DARTs (Directed Activity Related to Text), reporting, small group work, peer led discussions and puzzles. Some of these methods have been tried with remarkable success with second language learners. They prove highly motivating precisely because of their contextualised nature.

Conclusion

This chapter has explored the issue of language learning in science as it relates to second language learners. One objective was to bring out the complexity of the language question and the extent to which it is intricately woven into other issues like culture, cognition and educational background. The ideas of Vygotsky provide a useful theoretical basis for understanding these connections. His idea of language as a mediator between thought and action harmonise well with constructivist conceptions of creating structures which closely match what the learner already knows. Vygotsky views language as a tool that allows human beings to interpret and transform nature and self. Language thus changes with social and cultural change. As more second language speakers enter the English speech community, they also change the language itself. Given that English is the world's most widely spoken second language, the notion of a 'Standard English' is fast becoming something of the past. There is a dynamic interaction between the L2 learners of science and their teachers as English becomes ever more the lingua franca of scientific communication.

REFERENCES

Alvarez, A. A. (1991). Pilipino or English in science learning? The case of bilingual education in the Philippines. *Paper presented at the International conference on Bilingualism and National Development*. Brunei.

Allen, J. B. & Pitlanders, J. (1975) Adult Theories, Child Strategies and their implications in Clark, R. (Ed.) *Papers in applied linguistics*. London: Open University Press.

Bamgbose, A. (1984). Mother tongue medium and scholastic attainment in Nigeria. *Prospects XVI*(1): 87 - 93.

Beebe, L. M. (1981). Social and situational factors affecting the communication strategy of code switching. *International Journal of the Sociology of Language*, *47*:105-21.

Churms, S., Moore, R. & Paxton, M. (1995). Teaching future chemical engineers to write: UCT writing centre enters the laboratory milieu. In A. Hendricks (Ed), *Proceedings of the Third Annual Meeting of the Southern African Association for Research in Mathematics and Science Education*. Cape Town, South Africa: SAARMSE.

Cleghorn, A. (1992) Primary level science in Kenya: Constructing meaning through English and indigenous languages. *International Journal of Qualitative studies in Education*, *5*(4):311-323.

Cummins, J. (1980). The cross lingual dimensions of language proficiency: Implications for bilingual education and the optimal age issue. *TESOL Quarterly* 14:175-187.

Deckert, G.D. (1993). Perspectives on plagiarism from ESL students in Hong Kong. *Journal of Second Language Writing*, *2*(2):131-148.

Engel Clough, E., Driver, R. & Wood - Robinson, C. (1987).How do children's scientific ideas change over time? *School Science Review*. *69*(247):255 - 267.

Fairclough, N.(1989). *Language and power*. London: Longman.

Hewson, M. G. & Hamlyn, D. (1983) The influence of intellectual environment on conceptions of heat. Paper presented at the Annual meeting of the *American Educational Research Association*.

Hewson, M. G. & Hewson, P. W. (1983). Effect of instruction using student's prior knowledge and conceptual strategies on science learning. *Journal of Research in Science Teaching*, *20*(8):731-743.

Ingle R. & Turner A. (1981). Science curricula as cultural misfits. *European Journal of Science Education*, *3*(4): 357-371.

Inglis, M. (1993). An investigation of the interrelationship of proficiency in a second language and the understanding of scientific concepts. *Proceedings of the First Annual Meeting of the Southern African Association for Research in Mathematics Education*. Grahamstown, South Africa: Rhodes University.

Ivowi, U. M. O. (1984). Misconceptions in physics amongst Nigerian secondary school students, *Physics Education*, *19*(9): 761-767.

Ivowi, U. M. O. (1986). Students' misconceptions about conservation principles and fields. *Research in Science and Technological Education*, *4*(2): 127-137.

Janks, H. (1993). *Critical language awareness series*. Pretoria: Hodder and Stroughton and Wits University Press.

Kahn, M., Levy, S., Rollnick, M., Segall, R., & Kotecha, P. (n.d.). Some insights into the problems facing science education and possible policies to address them, *National Education Policy Initiative, Science Curriculum Group* report. Distributed as an occasional paper by Edunet, P.O. Box 1833, Houghton 2041.

Kahn, M. & Rollnick, M. S. (1993). Science education in the new South Africa, Reflections and visions. *International Journal of Science Education*, *15*(3):261-272.

Kay, S. (1975). Curriculum innovation and traditional culture a case study of Kenya. *Comparative Education* 11:183-191.
Khati, T. (1987). Language variability, code switching and communicative competence: Some preliminary remarks. Paper presented at the *Symposium of planning and coordination of Educational research*. Maseru, Lesotho.
Kotecha, P., Rutherford, M., & Starfield, S. (1990). Science, language or both? The development of team teaching approach to English for science and technology. *South African Journal of Education, 10*(3), 212-221.
Krashen, S. D. (1982). *Principles and practice in second language acquisition*. Oxford, UK: Permagon Institute of English.
Langhan, D. (1996). Roles of language and learning across the curriculum in learning and language across the curriculum *Draft Position paper*, The South African Applied Linguistics Association.
Lemke, J. L. (1989). The language of science teaching. In C. Emihovich, (Ed.), *Locating learning: Ethnographic perspectives on classroom research* (pp.216-239). Norwood, NJ: Ablex Publishing.
McNaught, C. (1991). Learning science at the interface between Zulu and English: A brief overview of research issues in this area. Paper delivered at a conference on *Language, Thought and Culture*. Johannesburg, 2-4 April.
McKinley, E., Mc Pherson ,Waiti P. & Bell, B. (1992). Language, culture and science. *International Journal of Science Education, 14*(3):579-595.
Moore, R. (1994). How do university students learn to write? In *Language and Academic Development at UCT*, Angelil Carter, S. Bond, D. Paxton M. And Thesen, L. (Eds). Monograph, Cape Town: University of Cape Town.
Nkopodi, N. & Rutherford, M. (1993). Misconceptions arising from language problems. Poster presented at the *First Annual Meeting of the Southern African Association for Research in Mathematics Education*. Grahamstown, South Africa: Rhodes University.
Ogawa, M. (1986). Towards a new rationale for science education in a non Western *society. European Journal of Science Education, 8*(2):113-119.
Ogunniyi, M.B. (1987).Conceptions of traditional cosmological ideas among literate and non literate Nigerians. *Journal of Research in Science Teaching, 24*(2): 107-117.
Opoku, J. U. (1983). Learning of English as a second language and the development of bilingual representational systems. *International Journal of Psychology, 18*: 271 - 283.
Poplack, S. (1994). Syntactic structure and social function of code switching. In Duran, R. P. (ed.) *Latino language and communicative behaviour*. Norwood, New Jersey: ABLEX.
Prophet, B. and Dow, P. (1994). Mother tongue language and concept development in science. *Language, Culture and Curriculum, 7*(3):205-216.
Raimes, A. (1991). Out of the woods: Emerging traditions in the teaching of writing. *TESOL Quarterly, 25*(3):407-430.
Rampal, A. (1992). A possible "Orality" for science. *Interchange, 23*(3):227-144.
Rohner, R. P. (1984). Towards a conception of culture for cross-cultural psychology. *Journal of cross-cultural Psychology, 15*(2):111-138.
Rollnick, M. S. (1988). Mother tongue instruction, intellectual environment and conceptual change strategies in the learning of science concepts in Swaziland. *Unpublished Phd Thesis*, Johannesburg: University of the Witwatersrand.
Rollnick, M. S. (1991). What strategies best help students who are learning chemistry through a second language? *Symposium presentation at the International Conference on Chemical Education*. York, United Kingdom.
Rollnick, M. S. (1994). The use of e-mail in Improving Student Writing and Concept Development. *Proceedings of the Southern African Association for Research in Science and Mathematics Education*. 469-478. Durban.
Rollnick, M. S. & Manyatsi, S. (1997). Language, culture or disadvantage: What is at the hear of student adjustment to tertiary chemistry ? Paper presented at the *Fifth Annual meeting of the Southern African Association for Research in Mathematics and Science Education (SAARMSE)* Johannesburg.
Rollnick, M. S. & Rutherford, M. (1996). The use of mother tongue and English in the learning and expression of science concepts: a classroom based study. *International Journal of Science Education, 18*(1): 91-104.
Rollnick, M. S., White, M & Dison, L. (1992). Integrating writing skills into the teaching of chemistry: an essay task with bridging students. *Proceedings of the Annual Conference of the South African*

Association for Academic Development. Port Elizabeth, 3-5 December.
Rollnick, M. S., White, M., Mphahlele, M. M., Giyose, A. & Bux, M.(1993). Developing students' writing through the presentation of posters. Paper presented at the Annual conference of the *South African Association for Academic Development*. Cape Town, South Africa.
Rosenthal, J.W.(1996). *Teaching science to language minority students* Clevedon: Mulitllingual matters.
Rutherford, M. (1993). Making scientific language accessible to science learners. Paper presented at the *First Annual meeting of the Southern African Association for Research in Mathematics and Science Education (SAARMSE):* Grahamstown, South Africa.
Sawyerr, E. S. (1979) The role of traditional beliefs in the teaching and learning of science in Sierra Leone. *Science Education, 22*(3/4).
Sinclair de Zwart, H. (1972). Developmental linguistics. In P. Adams (Ed). *Language and thinking*. New York: Penguin.
Solomon, J. (1983). Messy, contradictory and obstinately persistent. A study of children's out of school ideas about energy. *School Science Review, 65*(231): 225-229.
Solomon, J. (1984). Prompts, cues and discrimination. The utilisation of two separate knowledge systems. *European Journal of Science Education, 6*(3):277-284.
Stavy, R. & Wax, N. (1992). Language and children's conceptions of plants as living things. Paper presented at the *Annual meeting of the National Association for Research in Science Teaching*. Boston.
Strevens, P. (1980). *Teaching English as an international language* Oxford: Permagon Press.
Sutman, F. X. & Guzman, A. (1992). Teaching and learning for understanding to limited English proficient students: Excellence through Reform. *Eric Clearing House on Urban Education* New York.
Swain, M. & Cummins, J. (1979). Bilingualism and cognitive functioning and education in language teaching. *Linguistic Abstracts, 12*(1), 4-18.
Thijs, G. D. (1986). *Conceptions of force and movement. A study of Intuitive ideas of pupils in Zimbabwe in comparison with findings from other countries*. VAV report no.5, Free University of Amsterdam.
Vygotsky, L. S. (1978). *Mind in society: The development of higher psychological processes*. Cambridge, MA: Harvard University Press
Watson, P. & Bashe, L. (1996). *Voices from the classroom* Johanesburg: Sethlare Trust.
Whorf, B. (1956). *Language, thought and reality*. Cambridge, MA: MIT Press.
Wilkinson, A. (1982). *Language and education*. London: Open University Press.

[illegible] *Association for Academic Development*. Port Elizabeth, 3-5 [illegible].
Re[illegible], [illegible], White, M., Maphalala, M. M., Glyoco, A. & Bon, M. (1995). Developing [illegible] through the [illegible] of [illegible]. Paper presented at the Annual Conference of the *South African Association for Academic Development*. Cape Town, South Africa.
Rosenthal, J. W. (1996). *Teaching science to language minority students*. Clevedon: Multilingual Matters.
Rutherford, M. (1993). Making science the language [illegible] access [illegible]. Paper presented at the first [illegible] of the *Southern African Association for Research in Mathematics and Science Education* 54 [illegible], [illegible], South Africa.
Sawyer, E. S. (1979). The role of traditional beliefs in the teaching and learning of science in Sierra Leone. *Science Education*, 77(3-4).
Sinclair de Zwart, H. (1972). Developmental psycholinguistics. In P. Adams (Ed.), *Language and thinking*. New York: Penguin.
Solomon, J. (1983). Messy, contradictory and obstinately persistent: A study of children's out of school ideas about energy. *School Science Review*, 65(231), 225-229.
Solomon, J. (1984). Prompts, cues and discrimination: The utilization of two separate knowledge systems. *European Journal of Science Education*, 6(3), 277-284.
[illegible] & [illegible], N. (1992). Language and [illegible] a [illegible] of [illegible] learning [illegible]. Paper presented at the Annual Meeting of the National Association for Research in Science Teaching, Boston.
Stevens, R. (1986). *Teaching English as an international language*. Oxford: Pergamon Press.
[illegible] & Chatman, [illegible] (1987). Language and learning for understanding science in limited English proficient students. Excellence through Reform [illegible]. *Clearing House on [illegible]*. New York.
[illegible], M. & [illegible], J. (1979). Bilingualism and cognitive functioning and education in language teaching. *Linguistic Abstracts*, 2(1), 9-18.
[illegible], U. D. (1986). *Conceptions of force and movement [illegible] of [illegible] pupils in Zimbabwe in comparison with findings from other countries*. [illegible] report no 5. [illegible] University of [illegible].
[illegible], L. S. (1978). *Mind in society: The development of higher psychological processes*. Cambridge, MA: Harvard University Press.
Watson, [illegible] & Pastre, L. (1990). *Letters from the classroom*. Johannesburg: Sached Trust.
Whorf, B. (1956). *Language, thought and reality*. Cambridge, MA: MIT Press.
Wilkinson, A. (1982). *Language and education*. London: Oxford University Press.

Masakata Ogawa

Chapter 7

A Cultural History of Science Education in Japan: an Epic Description

Our ancestors found the one and only possible way to fight against a strong power of foreign culture: That is, ignoring thoroughly the fact that there existed a different principle in that culture from that of ours, and ignoring so perfectly as to be even unaware of the fact that they were ignoring it. By doing so, without fighting, they won the fight, with which other people could have struggled.
(Hasegawa 1986, p.52: Tr. by author)

Most research so far on science education in cultural contexts has been about negative effects of traditional culture on learners in non-western countries or on ethnic minority groups in western countries with respect to their learning science (see Wilson, 1981). Recent efforts on cultural studies in science education research have developed a new trend in which science itself is regarded as a kind of culture, for example, "science as culture" (Horton, 1967; Elkana, 1971 & 1981), "science as a worldview" (Cobern, 1991), "multicultural science education" (Hodson, 1993), "science as a culture of the scientific community" (Ogawa, 1995a), and "science education in a multiscience perspective" (Ogawa, 1995b).

This chapter adopts a different approach.[73] The fundamental presupposition is that both science and science education as they exist in any culture are interpretations or constructions by the people of a given culture. Though the context-dependency or culture-dependency of science education is well recognized in science education, context-dependency or culture-dependency of science itself is not so easily recognized even in the constructivist research tradition. That is, even when one refers to context-dependency of learners' knowledge construction, the context-independent existence of "science" is usually presupposed. In actuality, what one can access is not "science" in some brute form, but an image of "science" – and all images are inherently context-dependent or culture-dependent. For example, what Japanese people call "science" is an interpretation that integrates information from various sources available to the Japanese. "Science" is a constructed image that the Japanese believe to be culturally independent. And they even believe that their

[73] This chapter was originally developed as part of the 1996 International Joint Project on the Effects of Traditional Cosmology on Science Education (Project No. 08044003) supported by a Grant-in Aid for Scientific Research from the Ministry of Education, Science, Sports and Culture of Japan.

W. W. Cobern (ed.), Socio-Cultural Perspectives on Science Education, 139–161.

image of "science" is universal for other peoples. Science education, subsequently, is also an interpretation based both on the Japanese image of science and what the Japanese think science education should be. Thus, both images are closely linked to each other. This image construction is true for people in other cultures including westerners. Even scientists interpret the nature of science and construct images of science. There is no culture-free interpretation of science or science education. In this sense, there are many possible images of science and of science education each of which is inevitably dependent upon the particular culture concerned. Science educators as well as general public, however, are typically little aware of this difference among images, and believe that their image is the "real" science or science education.

The idea of the context-dependency of science and science education can guide us to a new cultural perspective in the enterprise of science education. In this chapter, I will try to describe the image of science, science education and their interrelationship within cultural historical context of Japan. If the description is done successfully and helps to illuminate the nature of science education, readers may find it worth their while to attempt a similar cultural-historical analysis of science and science education in their own countries.

Methodological Considerations

In this chapter, my use of the concept "image" is to convey the fundamental presupposition that one cannot directly access or interact with reality itself. What a person can actually access or interact with is an image of reality. The process of image construction is deeply buried under the cultural webs of meaning. In other words, the constructed image is deeply dependent upon one's culture. Reality involves various levels of objects: from concrete substances like trees, cats, water, the sun, and buildings, to abstract objects like meanings, symbols, concepts, theories, and worldview. For example, imagine a conversation between an Englishman, Mr. Smith and a Japanese lady, Ms Suzuki. They are talking about a pretty cat in front of them. Mr. Smith's word "cat" reflects his huge set of background knowledge, impressions, values and feelings. All of these are grounded in his web of meaning that is of course strictly dependent upon his culture. The same is the case with Ms Suzuki. Thus, Mr. Smith upon hearing Ms Smith speak the word "cat" cannot help interpreting her meaning through his own web of meaning. Never can he do this directly through *her* web of meaning. If he wants to know what she really means, he must try to construct an image of the word "cat" from her web of meaning. If the two live in a similar cultural/social situation, the gap between the webs of meaning may be small. Even in such cases, however, there is no guarantee that their respective webs of meaning are similar. Much the same can be said about science and science education. Using the key concept "image" is to declare that we do not adopt the position that there exists mutually independent subjects and objects, but do adopt the position that one should attend to interrelationship between the two. Separated descriptions of the subject and the object simply do not make sense. My approach is thus a "thick description" of science education and those involved with science education similar to research approaches in ecological anthropology, cultural history, and cultural geography.

This approach is not without precedence in other fields of inquiry. Ecological anthropologist Irimoto (1996) insisted that anthropology promote the analysis of a dynamic interrelationship between nature (ecology, living things) and culture (religion, society). He urged that the dualism of "nature versus culture" that was fundamental in structural anthropology and Marxist materialistic anthropology, be rejected. For Irimoto human activities and human behavioral strategies are at the center of interest and the interrelationship between nature and culture must be analyzed from that center in what he calls the "natural history of culture". Similarly, the anthropologist, Shimada (1993) proposes a "situational anthropology." Shimada turned from his background in Western and Eastern philosophy to African studies. His field work in the Cameroon introduced him to a world which is neither Western nor Eastern, and convinced him that human thought should be understood in the context of the daily lives people live rather than by idealized models. Kitoh (1996), an environmental philosopher interested in environmental protection, provides a third example. He too struggled with the traditional dualism between human beings and nature, and came to preference for a focus on interrelationship between human beings and nature. For this purpose he adopted the Japanese concept, "Seigyo," as an analytical framework. Seigyo refers to the human activities, by which daily lives are lived, within their natural environments. When focusing upon the Seigyo there is no boundary between humans and nature nor between culture and nature. Seigyo draws together the natural environment of an area and the culture (indigenous religion, festivals, symbols, food preferences, taboos, etc.) of the people in that area.

Following the precedent found in these examples, this chapter provides a *thick* cultural and historical background of the interrelationship between the Japanese people and science and science education in Japan. The chapter offers an *epic description* of science and science education in Japan in the hope that readers will be able to "share", "experience", and "live vividly in" the contexts described. However, readers are warned of a possible misunderstanding. Some may find a superficial resemblance between the Japanese science education described in the chapter and the science education of their own familiar cultural contexts, and think that the description is not in fact specifically limited to the Japanese context. My intention is not to insist that what the Japanese *do* in science education is unique to Japan. There are similarities with other cultures. Rather, I wish to show how current Japanese pedagogical practice is grounded in the historical experiences and changes of Japanese culture. I wish to show that in spite of surface similarities, many aspects of science education in Japan are quintessentially Japanese. Thus, I will try to offer a thick description of how the Japanese have faced, perceived, interacted with, and interpreted Western science and science education throughout Japanese history by taking an in-depth look at a few critical points in that history. This story begins with the Japanese people confronting their surrounding natural environment in Japan and foreign cultures coming to Japan.

The Japanese Way of Thinking

I begin with the Japanese way of thinking. Nakamura Hajime is a Japanese philosopher whose interest is the comparative study of thinking among Eastern peoples. In 1964, he published a seminal monograph, *Ways of Thinking of Eastern Peoples: India-China-Tibet-Japan.* The following discussion is based on his analysis of the characteristics of Japanese ways of thinking. Nakamura (1964, p. 5) uses the term, "ways of thinking" to mean, "any individual's thinking in which the characteristic features of the thinking habits of the culture to which he belongs are revealed." Nakamura (1964, p. 5) goes on to argue that the "thinker need not himself be aware of [these] ways of thinking when he is engaged in operations of thinking.... However, his ways of thinking are, in fact, conditioned by his culture's habits and attitudes when he communicates his thoughts."

Among many characteristics of thinking identified by Nakamura among the Japanese, I would like to identify those relevant to scientific thinking, communication about natural environments, and the acceptance of foreign cultures. These are as follows (the page numbers are from Nakamura, 1964):

> The Japanese are willing to accept the phenomenal world as Absolute because of their disposition to lay a greater emphasis upon intuitive sensible concrete events, rather than upon universals. (p. 350)
>
> The way of thinking that recognizes absolute significance in the phenomenal world seems to be culturally associated with the Japanese traditional love of nature. (p. 355)
>
> The Japanese love of nature is linked with the Japanese inclination to cherish minute things and to treasure delicate things. The tender love of animals traditionally runs in the veins of the Japanese, but that love is concentrated on minute lovable living things. (p. 356-357)
>
> The Japanese, owing to the tolerant and more open side of their nature, assimilated some aspects of foreign culture without much repercussion. They try to recognize the value of different cultural elements, at the same time endeavoring to preserve the values inherited from their own past. (p. 400)
>
> There is little intention to make each man's understanding universal or logical. In general, the thinking of most Japanese tends to be intuitive and emotional. (p. 531)
>
> The expressive forms of Japanese sentences put more emphasis upon emotive factors than on cognitive factors. It is extremely difficult to express abstract concepts solely in original Japanese words. (p. 531-532)
>
> Logical consciousness begins with consciousness of the relation between the particular and the universal; and the Japanese on the whole have not been fully aware of this relation, or have been poor in understanding a concept apart from particular instances. (p. 535-536)
>
> The historical neglect of logic is one of the salient features of traditional Japanese thinking. (p. 543)
>
> The Japanese people have seldom confronted objective reality as something sharply distinguished from the knowing subject. (p. 575)

These characteristics tacitly influence Japanese science education in various ways. They have certainly affected way Japanese people face natural environments and their way of life throughout their long history. They influence how the Japanese face foreign cultures, especially how the Japanese faced Western culture during the mid 19th century – Western systems of education and Western science. In a sense these nine statements summarize the Japanese characteristics that will emerge in this chapter.

Japan is a set of islands and, historically, a far peripheral zone with respect to two cultural giants. First it was peripheral to the ancient Chinese civilization and then later to modern Western civilization. The Japanese seem always to have faced overwhelming cultural pressures. Despite this history, the Japanese have maintained their cultural identity. The paradox is that Japanese culture through its long history has been able to adopt various components of foreign culture without losing its own identity. This is an important point when thinking about the inter-relationship between science education and the Japanese as well as between science and the Japanese.

Kosakai (1996) refers to this paradox as the "mystery of 'Cultural Openness' and 'Social Closedness.'" He argues that it is this *closedness* of Japanese society that in fact allows *cultural openness* among the Japanese. The Japanese are freely accepting of foreign culture, but the foreign culture is inevitably modified and transformed as it is being accepted. This modification and transformation is in terms of what the Japanese want the foreign elements to be like. Why does this kind of modification or transformation happen among the Japanese? Perhaps it is their historic isolation. Japan has never been inundated by a foreign culture; most Japanese have never lived in another culture. Their exposure to foreign culture is by importation of components.

Hasegawa (1986) argues from a language viewpoint that the Japanese traditionally learn about foreign cultures not through foreign languages but through the Japanese language. The oldest and most important example of the importation of foreign culture was the 5th century import of Chinese writing characters. The Japanese imported the Chinese characters as a tool for expressing the Japanese language but did not import the Chinese language. Hasegawa (1986) argued that the Japanese of the 5th century avoided losing the spirit or identity of Japan through this process, which would have happened had the language been imported. Later the Japanese wanted to learn many things from China, the cultural giant of East Asia at that time. Japan, thus, imported many documents written in Chinese; however, the Japanese did not try to read and understand the documents in the original Chinese language. The meaning was always interpreted through their own Japanese language. For the purpose of this process, the Japanese invented a systematic methodology, in which the sentence of original Chinese was never read in its own pronunciation nor understood by itself. The original order of Chinese characters was changed so that the meaning of the sentence was interpretable in Japanese. The way of the Japanese is to learn foreign culture through the Japanese language (Hasegawa, 1986). An important consideration is that the Japanese were never aware that they were in fact ignoring the essence of the original culture. They understood themselves to be doing just the opposite. This attitude is found in Fukuzawa Yukichi, one of the key persons responsible for the acceptance of modern Western culture at the beginning of Meiji era. He translated many Western books into Japanese in order to import Western culture. His philosophy of translation was "translation with daily words and daily expressions" (Anzai 1995, p. 13-17) that could be easily understood by the average Japanese person. This philosophy inevitably ran into difficulties as Fukuzawa tried to explain the *original* Western meanings in everyday Japanese language, especially when it came to abstract ideas such as those typically found in Western science. The unintended consequence of translation was that Japanese words used in translation took on meaning of their own. These new meanings could differ with both the Western meanings and the traditional Japanese meaning of the works. The point of significance is that this tradition is still alive in Japan. One can even say that this is the only way that the Japanese learn anything about foreign cultures including foreign languages.

Similarly, with regard to science education, from elementary school to graduate school, the instructional language for science is Japanese. This suggests that the essential nature of science as defined by European languages is effectively masked in Japanese schools. When European language based science is translated into Japanese, terms are borrowed from everyday Japanese language. Inevitably what at first would seem to be a "scientific" term carries with it the original meaning of the Japanese word but in a hidden form. New scientific meaning and the original meaning are mixed giving rise to what might be called "Japanized" scientific meaning. Other terms must be newly invented and this is done by combining Chinese characters to form an original term. In the Japanese language system, new terms can be easily invented to express new ideas because a new combination of Chinese characters (each of which has its own meaning in itself) signifies a new meaning. Still other new terms can be expressed in *Katakanas*. Katakanas is another character system for expressing the Japanese language but in this case the characters have no individual meaning. Definitions are paraphrased and described but since this is done within Japanese culture, Katakanas expressions always bear the influence of Japanese culture. Again, the expression of science in Japanese "Japanized" and this situation raises interesting complications for science education.

The Roots of Japanese Education

Prior to the Meiji Restoration in 1868, Japanese literacy under the Tokugawa regime, was comparable or better than that of Western peoples of the period. Kobayashi (1976, p.13, p. 18) estimated that:

> toward the end of the Tokugawa period there were some 1200 Terakoya in and around Edo [present day Tokyo] and that the total number in the country was about 15,500.... (F)or the commoners, it is generally agreed that about 40 to 50 percent of the male population and about 15 percent of the female population were literate or had received formal education.

Terakoya were private schools run by public-spirited individuals (Murthy, 1973). Other estimates place the percentage of adult male literacy even higher. Oishi and Nakane (1986) estimated that about 80% of the male population in Kyoto were literate. By comparison Cipolla (1969) estimated that during the mid 1800s adult literacy in Denmark, Finland, Sweden, Norway, Scotland, Germany, Switzerland, and Netherlands was about 70%. In Austria, England, Wales, France, and Belgium it was 50-70%; and in Russia, Italy, Spain, Hungary, Greece, Rumania, and Portugal the adult literacy rate was less than 50%.

During the 19th century the Japanese cities of Edo, Osaka, and Kyoto were among the largest cities in the world. These cities had many Terakoya schools where the children of commoners went to learn basic literacy and numeracy. Even in the rural areas, however, there were Terakoya schools for the children of farmers. These schools were established in response to a growing need for literacy among commoners. Literacy was important in 19th century Japan for several reasons. Government orders and commands were written on public notice boards. Japan had a monetary economy system that meant farmers had to buy fertilizers, for example, even for even small fields. Literacy was essential for commercial activities.

Moreover, parents of commoners wanted their children to be literate because that would increase a child's chances of getting work in the cities. This was desirable because one could earn more money. Also education was highly valued in the Confucian system (Tada, 1992). The poor were able to attend the Terakoya schools because there was no formal tuition fee system. Most Terakoya teachers were samurai warriors, priests, or medical doctors all of whom had their own income. Serving as a Terakoya teacher was a service to the community. The result was that even poor farmers could send children to school. Instead of a fee, farmers sent rice or vegetables, not as a fee for tuition, but a kind of gift expressing appreciation.

A modern school system came in Japan in 1872 with the establishment of Gakusei (The Fundamental Code of Education) and Shogaku Kyosoku (Regulations of Elementary School). Government leaders at that time decided to import European and American school system into Japan because they saw this as necessary for attaining the Western standard in industry, technology, wealth and power. Terakoya schools were of course thoroughly Japanese with thoroughly Japanese curricula. With Gakusei and Shogaku Kyosoku new concepts were introduced and thus new challenges.

Shizen and Nature

The new concept of interest brought by the new curricula is "nature". It is of interest because it is so central to science and science education and because until the late 19th century, there was no concept in Japan corresponding to this Western concept. When the concept of nature appeared in Japanese curricula it was (and remains so today) by way of the Japanese concept of :Shizen."

Shizen is an old Japanese word originally from the Chinese word, "tzujan." Kawasaki (1990) argues that the original meaning of "tzujan" is state of spontaneity and which stands for the highest virtue. According to the ancient Chinese philosopher Lao-Tzu, all things ultimately come out of "tzujan." The original meaning of Shizen is "onozukara-shikaru," that is, "of itself", or "spontaneously." Shizen is an adverb, in fact, not a noun in this traditional meaning which is still part of modern Japanese usage. Shizen has a positive image with the Japanese in the sense, for example, that "Shizen is the ideal of what everything should be" (Watanabe, 1992, p. 179). One may notice similarities with the English concept of nature. For example, an English native might say that "Nature is the ideal of what everything should be." There are, however, greater differences between the Japanese and English concepts. In Japanese, there is no single word for capturing the thought of natural things and natural phenomena in total the way that "nature" does in English. There are expressions in which the names of concrete natural things are set in a phrase, for example, "San sen sou moku" (literally: mountain, river, grass, tree) and "Sou moku chu gyo" (literally: grass, tree, insect, fish) but these expressions never stand for natural things in general as such are understood in English. To better under-stand the Japanese sense of Shizen, one must understand the Japanese sense of their environment.

Shimada (1993) argues that the world that contemporary Japanese call Shizen may just as well be called "Fudo" following the usage of Watsuji Tetsuro, one of the great thinkers of Japan. Shimada (1993) explains that Watsuji's usage of "Fudo" is not merely for natural environments but for the interrelationship between human activity and the natural environments in which that activity occurs. The two should be understood as a whole, or as "Fudo." In other words, for the Japanese their perception of natural environments and activities within natural environments are not separable from natural environments. It is the sense of interrelationship that is fundamental and this sense is rooted in both the geography and history of Japan.

Japan is a series of narrow islands at the east edge of the Asian continent. There is a series of high mountains (about 3000-2000 m) at the center of the islands, and at the foot of them there are small mountains and hills covered with mainly broad-leafed forests. Flatlands are usually narrow and located at the seashores or in basins surrounded by the small mountains and hills. The climate is basically under the influence of monsoons with rainfalls at about 1700 mm/year. There are four distinct seasons, spring, summer, autumn, and winter, but the weather frequently changes drastically over a period of only a few days. One of the characteristics of the climate is that there is a rainy season between spring and summer. Heavy rains, sometimes 300-400 mm/day, can come for several days. In ancient Japan there were no embankment works, thus floods happened almost every year. The typhoon season comes to Japan every year at the end of summer and early autumn. The Japanese islands also hold a significant number of volcanoes and earth-quakes are frequent. The rivers are relatively short (even the longest rivers are only 350-250 km in length) and have steep slopes. The natural environment of Japan is never monotonous.

The earliest of Japanese lived during Jomon era of 12000BC to 300BC. Recent research findings have added detail to what is known about life in the Jomon era (Sasaki 1991). After 9000BC, as the climate warmed, deciduous broad-leafed forests began to dominate Japan. Jomon people lived a rich settled life in the flatlands near the fruitful deciduous broad leafed forests and seashores. They eat wild plants and acorns, nuts and various kinds of wild root crops collected from the forests and flatlands. In spring they collected shellfish. In summer they could fish. In autumn they both collected acorns, nuts and wild root crops, and fished for salmon in the rivers. Food was stored for the winter. At a very early date the Jomon had primitive techniques for food processing. In addition, there is evidence of small-scale primitive cultivation and incipient agriculture. After 6000BC, further warming allowed evergreen broad leafed forests to dominant the Western and southern parts of Japan which led to further changes of lifestyle now based on swidden (or slash-and-burn) cultivation (Sasaki, 1988). People could supply most food needs from such agriculture; but productivity was still not so stable and collecting acorns, nuts, wild root crops, hunting, and fishing remained important. Jomon culture was fruitful and Yasuda (1993) called it the "Civilization of the Forests." Life in the Jomon era, however, was heavily dependent upon the natural environment and climate. At its best, the natural environment could be "merciful mother." It could also be the "stern father" visiting the Jomon with floods, typhoons, snow, droughts, earthquakes, tidal waves, and volcanic eruptions. In response the Jomon developed a religious system of animism. Everything – animals, insects, birds, plants, nonliving things, and natural phenomena like thunder, and weather – has its own spirit. The spiritual and the natural were intertwined.

The Yayoi era in Japan began after 300BC and continued until 300AD. During this period the primitive agriculture of the Jomon period was gradually replaced by the sophisticated and systematic cultivation of rice. Again changes in climate – this time increased cold spells – brought changes of lifestyle. Rice agriculture did not spread to the central and eastern parts of Japan because there the deciduous broad-leafed forests protected against the coldness and people continued to live off the forests. There the Jomon culture survived. In the flatlands, however, the broad leaf evergreens and other vegetation suffered with the cold. Food sources were destabilized. When migrants to the flatlands introduced the new technology of rice agriculture the indigenous people readily accepted the new technology. The rapid spread of rice cultivation was considerably helped by compatibility with the existing practices of swidden cultivation. Thus, two peoples were gradually mixed and from this came the modern Japanese – a people upon whose hearts and spirit are the deeply impressed scenes of rice fields surrounded by evergreen broad leafed forests.

Rice cultivation enabled people to have more food than swidden cultivation alone could provide. Rice was stored for the winter seasons, the people could afford more children, and the population grew. Communities were established and managers of communities were differentiated by the status of ruler. Surplus rice served as "money" and as society prospered, there appeared rulers who did not have to engage in rice production. These now had time to contemplate the spiritual face of natural things and they cultivated a sense of respect, a special feeling for natural things. In turn, they taught others this respect as the concept of "one-body-ness." This simply means that human beings and every natural thing are one body in total. The orientation of totality appears in people's minds as the feeling of love for natural things just as if the natural things were the people themselves. The point I wish to stress is that the modern day Japanese spirit and way of life associated with the concept of "Shizen" is deeply rooted in the Jomon, "Civilization of the Forests," culture and the Yayoi culture of ancient Japan.[74]

As noted above, the coming of Western science to Japan brought a concept of nature considerably at odds with the Japanese concept of Shizen though eventually the two were to become linked in Japanese school science.

The Encounter with Western Modern Science and Science Education

By 1543, the first piece of Western technology, the gun, had arrived in Japan. Later, further technology was brought by the Jesuits. Because Japan was at that time a literate society, the Japanese were able to utilize the imported technology for Japan's own purposes. In the 17th century, the Tokugawa regime began to restrict Western contacts because of the threat that Christianity posed for the traditional religions of Japan. The Tokugawa regime, however, maintained trade contacts with the Dutch and Chinese (who had no religious ambitions) through the harbor at Nagasaki. Western technology and science continued to enter Japan through books written in Dutch or by Chinese translations of Western books. "Western studies" became an important discipline for scholars who subsequently published translated versions of the Dutch and Chinese books as well as original works on science and technology.

---

[74] This is amply demonstrated through recent systematic research in archeology, genetics, ecology, folklore, and anthropology (see, e.g., Sasaki & Matsuyama, 1988).

Thus, throughout the 16th and 17th centuries, Japan became increasingly knowledgeable of Western science and technology. Note that the Japanese received Western technology and science through the medium of books, and not by close, long and continuous contact with Westerners. Kosakai (1996) insists that Japan is a unique society in this regard. Historically the Japanese experienced relatively few direct contacts with foreigners until quite recently. In a sense, the Japanese contact with foreign cultures has mostly been indirect through translation.

The Meiji Restoration in Japan, which was of significant intellectual and cultural importance, came in 1868. It is important to note that the mid 19th century was also a period of great intellectual and cultural importance in Europe. Science was becoming socially recognized and professionalized (Iyama, 1989, Murakami, 1994). For example, the British Association for the Advancement of Science was founded in 1831. Whewell, in his 1840 monograph, *The Philosophy of the Inductive Sciences*, coined the word, "scientist." Thus, Westerners and the Japanese met modern science in its initial developmental stage contemporaneously and at almost the same period of cultural and technological development. Much greater differences lay in the qualitative content of culture (as in the difference between "nature" in European culture and "Shizen" in Japanese culture) and school curricula.

In the 1860s, for example, science related subjects began to appear in European and American secondary school curricula, and after the 1870s, in the elementary school curricula (Turner 1927). During the Meiji era Japan sent many students to study in the USA and Germany, and the development of Japanese education was subsequently influenced by student experiences with "Pestalozzian object teaching" and Scheldon's Oswego movement. These were of course part of the "nature study" movement begun in the late 1880s, the spirit of which was the emphasis of "nature-sympathy and the nature-relations" (Bailey 1911, p. 17). In nature study, science was not a separate subject but taught in conjunction with natural history and natural philosophy, history, and geography (Iwasaki, 1977). This was particularly true in German (i.e., Prussian) elementary education after the institution of the 1872 law, Allgemeine Verfugung uber Einrichtung, Aufgabe und Ziel der Preussischen Volksschule. This law was in effect for about 50 years and as a result Japanese students of European educational practice were much influenced. The similarities between the Western concept of nature study and the Japanese concept of Shizen are not difficult to see. They both stress interrelationship. Indeed, it was in the 1890s that Shizen emerged from among several Japanese words[75] as the popular choice for translating "nature." Yanabu (1982) believes that Mori Ogai, a famous writer as well as a doctor trained in Western medicine, was the first to translate nature as Shizen" in the context of science. Shizen thus acquired new meaning as a noun but the concept clearly retained its Japanese system of cultural reference.

The Development of *Rika* - Japanized Science Education

As discussed above, prior to the Meiji era Terakoya schools provided, in the voluntary and informal setting, the education for most of the Japanese who were able to attend school. The curricula of Terakoya included reading, writing, and arithmetic,

[75] Up until this time a number of different Japanese words at different times had been used to translate the English, "nature," for example, "Tenchi" (heaven and earth) and Ban'yu" (universe).

with the addition of sewing for girls. The year 1872, when the Shogaku Kyosoku (Elementary School Curriculum) was established, was the very beginning of the Japanese formal elementary school system. The Shogaku Kyosoku brought the study of natural phenomena into the school curricula for the first time. The components were: Yojo Kuju (hygiene[76]), Kyurigaku Rinko (natural philosophy), Hakubutsu (natural history), Kagaku (chemistry), and Seiri (physiology). Itakura (1968) estimated that 14% of the total classes were devoted to such subjects. Although the elementary schools were intended to replace the Terakoya schools, they were not always successful. The enrollment rate for elementary schools in 1873 dropped to only 28%. Moreover, many pupils left school within a few years because parents could neither afford the costs of education nor afford the loss of income occasioned by children who attended school rather than work. Unlike the Terakoya schools, the elementary schools involved fees and operated on schedules not as compatible with traditional Japanese life.

Kyoikurei (Education Act) and Kaitei Kyoikurei (Revised Education Act) were enacted in 1879 and 1880 as further reforms. These reforms divided elementary education into three levels: primary level for three years, middle level for three years, and upper level for two years. Only the primary level was compulsory for all pupils and at this level there were no subjects on natural phenomena. At the middle level, students studied Hakubutsu (natural history) and Butsuri (natural philosophy). At the upper level, the subjects included Kagaku (chemistry), Seiri (physiology), Hakubutsu (natural history), and Chimon (physiography).[77] Where textbooks were used they were usually translated versions of Western textbooks. Students read factual descriptions of natural phenomena with little evidence of the influence of Japanese traditional culture. Itakura (1968) estimates that only 8.6% of total classes were devoted to such subjects. While the enrollment rates at the primary level during 1881-84 were at 40-50%, only 2-3% completed the middle level and only 0.2-0.4% completed the upper level where the science related courses were offered.

Yet another series of reforms came in 1886 when the government enacted Gakkorei (School Regulation Law) which included Shogakkorei (Regulation Law of Elementary Schools). The elementary school system was divided into two levels: a compulsory normal level (grades 1-4), and an optional advanced level (grades 5-8). According to the Shogakkorei, Shogakko-no-Gakka-oyobi-sono-Teido (Elementary School Curriculum and Its Level) was issued in the same year. There were no subjects related to natural phenomena at the normal level. At the advanced level (5th to 8th grade), however, there was a new name for the subjects related to natural phenomena, *Rika*. Indeed, Rika was a new name for subjects very similar to the old subjects of Hakubutsu, Butsuri, Kagaku, and Seiri in the Shogakko Kyosoku Koryo. Rika classes met twice a week in each of the four years of the advanced level so there was actually a decrease in the amount of time given to these subjects.

At that time Rika was a new word to most Japanese including most Japanese educators. The origin of the word Rika is not clear. Itakura (1968) suggested that it

---

[76] "Hygiene" courses were about personal health care and were patterned on contemporary American "health" education.

[77] The subject areas were set by the Kaitei Kyoikurei, and Shogakko Kyosoku Koryo (Curriculum Guideline of Elementary School) issued in 1881.

came from the Teikoku Daigakurei (Regulation Law of Imperial University) issued just before the Shogakkorei. This law included the Rika Daigaku (College of Rika) among other disciplines such as law, literature, medicine, and engineering. Rika was regarded as a collective subject consisting of natural history, natural philosophy (physics), chemistry and physiology:

> Rika treats things closely relative to pupils' lives like fruits, grain, vegetables, trees, grasses, human body, animals, insects, fish, gold, silver and iron, and things; and phenomena pupils can see in their daily life like sun, moon, stars, air, temperature, steam, clouds, dew, frost, snow, mist, ice, lightening, rain, wind, volcanoes, earthquakes, tide, burning, rust, rot, pump, fountain, sound, echo, watch, thermometer, barometer, steam instruments, eyeglasses, color, rainbow, lever, pulley, balance, magnet, and telegraph. (Shogakko-no-Gakka-oyobi-sono-Teido, Article 10, part Rika. Tr. by author)

This was only a topic list with no description of how such topics should be taught and little indication of cultural influence.

Monbusho (Ministry of Education) enacted one of the earliest documents prescribing the aims of Rika in elementary school, Shogakko Kyosoku Taiko (Syllabus of Elementary School), in 1891. It is important to carefully examine this document from the perspective of Japanese indigenous culture toward natural things and phenomena because the spirit of this document is alive today in Japan's elementary Rika program. The aims of Rika are given in Article 8 of the document, Shogakko Kyosoku Taiko (Syllabus of Elementary School):[78]

> Rika aims at making pupils: (1) observe (Kansatsu)[79] the usual natural things and phenomena precisely, (2) understand the outline of the interrelation among natural things and phenomena as well as the relation of such natural things to the pupil's lives, and (3) *cultivate the love of natural things*. In the lower grades level, the subject matter should be living things (plants, animals, birds, fish, reptiles, insects, worms, etc.) and minerals familiar to pupils and found around the school, and natural phenomena. Rika especially aims to have pupils observe the forms, lives and developmental processes of important living things and understand the outline of them. In the upper grade levels, teachers should make pupils understand the interrelation among living things as well as the relation of such things to the pupil's life, familiar physical and chemical phenomena, and the structure and mechanism of familiar instruments or apparatuses pupils can usually find, and also teach the outline of human physiology and hygiene. In the subject, Rika, teachers should teach those topics which are suitable for people's lives, like agriculture and industry. When teaching about living things the outline should include how to make important products from them and how such products are useful for people's lives. In the subject, Rika, it is important that teaching should be based upon observing (Kansatsu) the concrete objects, showing the types, models and drawings, or making simple experiments. (Tr. by author; emphasis added)

Most historians of Japanese science education have not thought that this document shows the "spirit of science." They argue that the document reflects the spirit of Jung's (1885) "Lebensgemeinschaft (life-community)" theory, which was imported from Germany at that time and reflected the rise of the Japanese nationalism or militarism at that time (Itakura, 1968).[80]

---

[78] Unfortunately, many governmental documents on educational policies pertaining to this period were lost in two disasters. The first was the Great Tokyo Earthquake and fire on September 1, 1923. The second was the Great Tokyo Air Raid of March 10, 1945.

[79] For an in-depth discussion of the Japanese concept of Kansatsu, see Kawasaki (1992).

[80] It is true that Japanese nationalism came into the educational policy in those days. The Kyoiku Chokugo (the Imperial Prescript on Education), which is regarded as the symbol of Japanese nationalism, was

Okamoto and Mori (1976) and Ogawa (1986b), however, dispute the claimed influence of Jung and argue instead that Rika actually shows the deep influence of the Japanese traditional view of Shizen.[81] They point to section 3 of Article 8 (quoted above) which specifically identifies the "cultivation of the love of natural things" as a Rika objective. This they say was no thoughtless addition given the make up of the committee that drafted this law (Shogakko Kyosoku Taiko). According to Itakura (1968), the 1891 drafting committee consisted of four members (Muraoka Han'ichi, Takamine Hideo, Nojiri Seiichi, and Shinoda Toshiei). All had studied abroad in order to learn more about Western education and at the time of their committee service were involved with Japan's Higher Normal Schools. Itakura (1968) suggests that Muraoka took the initiative because he was the sole committee member with a doctorate degree in science and he was, moreover, the eldest and this factor is crucial in traditional Japanese human relations. For several reasons, it is also very likely that Takamine Hideo contributed significantly to the drafting of Article 8. Takamine was one of Japan's leading educators; and, not unimportant, Takamine was only one-year younger than Muraoka. Takamine was also the principal of the Tokyo Normal School when Nojiri and Shinoda (the two other committee members) were students there. This would typically be a crucial factor in traditional Japanese human relations. Thus, it is important to take a closer look at both Muraoka Han'ichi and Takamine Hideo with regard to Article 8. Let me begin with Muraoka.

Muraoka Han'ichi was born in 1853 in a small village in Inaba (Tottori Prefecture). He was the son of a doctor trained in Western under Ogata Koan, a famous scholar of Western studies. Young Muraoka attended a Han school, Shogakukan[82]. Because of his superior achievement he was selected as a fellow and sent in 1870 to Daigaku Nanko, one of the national universities at that time. He studied German and physics and in 1878 was sent by Monbusho to Prussia. In Prussia he studied Prussian educational affairs and furthered his studies in physics at the Strasbourg Normal School and the University of Strasbourg. While in Prussia he translated into Japanese a German text on educational philosophy titled, *Die Praxis der Volksschule*[83] under the title, *Heimin Gakko Ronryaku*, in 1880. This was the first introduction of the philosophy and system of German education into Japan. At

---

issued in October in 1890. Japanese nationalism sought to promote Western technology while protecting Eastern morals.

> The traditional Kokugaku (the study of Japanese classics) and Confucian schools came back to the scene of national politics of education. The Occidentalists themselves began to take more the line of nationalistic thought. The Westernization of education was allowed to proceed only within the limits of technology and related practices, that is, the curriculum structure, teaching methods, school organization, etc. The "moral" aims of education were cautiously prescribed and interpreted through the traditional national philosophy, the most elaborated expression of which is to be found in the Imperial Prescript on Education of 1890. (Kobayashi, 1976, p. 29, Tr. by author)

Japanese nationalism, however, does not adequately account for the Shizen aspects of Rika.

[81] This is quite important issue when examining the interrelationship between the Japanese traditional view of Shizen and the school subject, Rika. This issue will be discussed in the later section, Contemporary Japanese Education.

[82] Han schools were the schools established in feudal domains or baronies.

[83] The author was C. Kehr who published the work in 1872.

this time Muraoka also studied physics at August Kundt's laboratory along side fellow student, Wilhelm Conrad Röntgen (1845-1923), who later was awarded the first Nobel prize in physics. Muraoka successfully completed his Ph.D. in physics in 1881, and was the first Japanese person to publish a paper in an international journal of physics. Thus, Muraoka had excellent credentials in Western science and on his return to Japan in 1881, Muraoka was appointed professor of physics at the Medical College of the University of Tokyo. In 1886 he was appointed head professor of Daiichi Senior Secondary School, Tokyo. In 1890 he was appointed head professor of the newly established Woman's Higher Normal School.

Dr Muraoka's views on science teaching in elementary school are clearly stated in his 1883 paper, *Butsurigaku Kyoiku Ho* (Teaching Method of Physics). Muraoka wrote:

> If you want to know how to teach physics in elementary schools, you should clarify the purpose of teaching physics.... Physics is a kind of academic discipline. However, investigating an academic discipline is not the direct purpose of elementary schooling. Physics can cultivate the basis of art or technology, and capability of research. Though this may be a kind of purpose of teaching physics, the purpose of teaching physics in the elementary school level should be to consider the daily phenomena that pupils face. In general, those who grew up without formal schooling (farmers and merchants in these days), could guess improbable reasons of daily phenomena, find difficulty to understand even trivial matters, and never know the usefulness of instruments like a barometer, thermometer, and telegraph. Originally, elementary school is the place to teach the minimum knowledge without which people cannot live in their community. (Muraoka 1883, Tr. by author)

Recalling Muraoka's choice to translate the German *Die Praxis der Volksschule* for use in Japan, one finds that this book clearly sets out a distinction between Naturlehre, or natural philosophy and Naturgeschichte, or natural history, with the latter promoted as the better subject for lower elementary grades. Examining the focus in the above passage on things relevant to pupils' lives, one finds similarity not only with the Article 8, but also with the following passages from *Heimin Gakko Ronryaku*:[84]

> Natural history treats the characteristics and phenomena of natural things, and classifies them, gaining an understandable perspective, according to the similarity of their essential characteristics. On the contrary, natural philosophy treats the uncovering the natural laws, under which all things, whether they are organic or inorganic, are governed. Natural history is a lighter subject for elementary school, but natural philosophy is a heavier one for elementary school. (Tr. by author)

With regard to section 3, Article 8, phrase, "cultivate the love of natural things" the following passage from *Heimin Gakko Ronryaku* is of even greater interest.

> Empirical, logical, *aesthetic* and religious interests are cultivated by Natur-geschichte (natural history) teaching in commoners' school.... Naturgeschichte is a discipline, not only to show us the infinite power of nature, but also what are useful or useless among natural things..... Naturgeschichte teaching should stimulate pupils' *aesthetic interest*. Beautiful colors and forms teacher presents impress pupils. Natural things pupils observe in their native place should give them *aesthetic feelings*. *Aesthetic experience* is one of the major aims of Naturgeschichte teaching. When teachers lead pupils to *aesthetic feelings*, *religious interest* appears among their respective minds. (Tr. by author, emphasis added)

---

[84] Muraoka's Japanese translation of *Die Praxis der Volksschule*

Of course the original author, Karl Kehr, wrote under the influence of the German spirit of nature, deeply influenced by German romanticism.[85] Muraoka, on the other hand, would have found a natural affinity or resonance between his own Japanese sense of Shizen and the German romantic views of nature - if only superficially. Kehr's "aesthetic feelings" become Muraoka's "cultivate the love of natural things" leading Muraoka to think that "love of nature" is a shared ideal in the traditions of both Japan and the science of the Western world. Thus, the wording of Article 8 – a science education policy document – becomes less problematic.

Takamine Hideo, the second critical member of the 1891 Shogakko Kyosoku Taiko drafting committee, was born in 1854 at Aizuwakamatsu, the first son of a samurai warrior. Takamine entered the Han school, Nisshinkan, at the age of eight and graduated at eleven. Since most other students were entering at ten and graduating at fifteen, he was called a genius. The curriculum at Nisshinkan was based upon Chinese studies. It is not clear that Western studies were included at Nisshinkan, but there are documents which show Western studies were introduced in the Aizuwakamatsu area during Takamine's school years (Ishikawa, 1902; Kaigo, 1971). Takamine later entered the Keio Gijukyu in Tokyo which was established by Fukuzawa Yukichi, in 1871. There he studied a more Western oriented curriculum of liberal arts and social sciences, history, geography, economics and arithmetic. He did not, however, study the natural sciences.

In 1875, Takamine was sent by the Meiji Government to the United States to investigate the American normal school system and there he studied at the State Normal School, Oswego, from which he graduated in 1877 (Murayama, 1978). This school was known for using Pestalozzian object teaching (Bailey, 1911). There he studied some of the natural sciences including physiology, zoology, botany, but not physics (Murayama, 1978). Of particluar interest is that Takamine studied zoology under Dr. and Mrs. Straight both of whom had been students of the famous, Louis Agassiz (Ahagon, 1989). Agassiz with a flair for the romantic was known for his motto, "Study nature, not books" (Bailey, 1991, p. 16). Ahagon (1989) also indicates that Takamine attended the 1877 Summer School of Salem which was conducted in the spirit of Agassiz's motto, "Study nature, not books," and the teaching method was patterned on Agassiz's (1886), *Teaching from Nature Herself*.

The key point here is that Louis Agassiz stressed the importance of observation in doing science. The English word "observe" is typically translated in Japanese as "Kansatsu." This translation, however, is no less problematic than the translation of "nature" as "Shizen." When the Japanese use Kansatsu with respect to natural objects and phenomenon, it is more than "observation" that takes place. The Japanese "gaze." They commit themselves to the activity and the object - which properly understood is not an object at all. In Kansatsu one is urged to experience "one-body-ness" (Kawasaki, 1992).[86] To the extent that Takamine interpreted the Western sense of subject/object observation as Kansatsu, he would have felt very

[85] German romanticism of the 16th-18th centuries is linked to the ancient cosmology of the "Great Chain of Being," in which every person is thought to be linked with every other person who ever lived. The cosmology originated with 3rd century Greek philosopher, Protinos, the founder of Neo-platonism (Lovejoy, 1936).

[86] According to Kawasaki (1992) Kan in Kansatsu contains the nuance that the subject performing the watching comes to the feeling of one-body-ness with what is being watched.

much at home with Western science. Moreover, Louis Agassiz's emotional stance toward nature would only confirm Takamine's interpretation. Thus, like Muraoka, Takamine had a peculiar experience in that his exposure to Western science was exposure to a variant of Western science atypically consonant with important aspects of Japanese culture, Shizen and Kansatsu. The influences of Takamine's American experiences are evident on his return to Japan. Takamine choose to translate for use in Japan James Johonnot's book, *Principles and Practice of Teaching* which drew heavily upon the concepts of the American nature study movement and the spirit of Louis Agassiz. On returning to Japan, Takamine served as professor at normal school and university levels, and championed the Pestalozzian movement and object teaching in Japan before joining Muraoka on the drafting committee of 1891.

With regard to Article 8, the importance of which should not be underestimated, we now have the background for the two principle framers. Muraoka and Takamine developed a notion of science that was eminently compatible with the traditional Japanese concepts of Kansatsu and Shizen, and a philosophy of "science" teaching for the elementary grades that emphasized the importance of an emotional sense of unity, the importance of everyday experience and everyday objects, and the importance of having pupils involved with their hands, that is, Rika. Rika thus reflects a view of education and culture that is as old as the "Civilization of the Forests" and yet as contemporary as the newest elementary school building in 20th century Japan.

Contemporary Japanese Education

Japanese education from kindergarten to upper secondary school is under the strict control of the central government and always in the Japanese language. Textbooks, though written by teams consisting of leading in-service or retired teachers, science supervisors, scientists, and science education professors and published by private publishers are authorized by the Ministry of Education, Science, Sports and Culture (Monbusho). Teachers are required to have teaching certificates which are issued by municipal authorities under the control of the national law which prescribes precise standards. National laws also prescribe the numbers of pupils per class, the number of school days, and even what scientific instruments and apparatus a school is to have. School curricula and course content for all subjects are legislated under Gakushu Shido Yoryo. Gakushu Shido Yoryo prescribes the course of study that teachers must follow and the authorized textbooks that must be used.[87] Pupils of the same age all over Japan learn virtually the same school subjects, with the same content, using the same materials and instruments or apparatus at the same time.

Rika is the subject intended to correspond with Western science. While Rika is not set for 1st and 2nd graders, for 3rd to 6th grades, Rika consists of one subject, simply called Rika which meets for three 45 minute classes a week, 35 weeks a year. At 7th to 9th grade, Rika is divided into two courses, one required, the other optional. The required course meets for three 50-minute Rika classes a week for three years. The first part of the requisite Rika course includes the physical and chemical sciences. The second part is mainly about the biological and earth sciences. The

[87] The option to use no textbook is not available to teachers. They must use textbooks and these must be the authorized books.

optional course, Sentaku-Kyoka Rika, is activity-oriented and meets once a week. It is offered, in addition to the required course, for 9th graders more interested in Rika than in other major subjects like Japanese language, mathematics, social studies, and English.

At the upper secondary level, Rika is divided into thirteen courses categorized by four groups. The first group of courses is oriented toward the everyday-life world and is mainly for students who do not want to go to college or university, or who will attend private colleges or universities to study subjects other than science and science related subjects. The examinations for these tertiary institutions do not include Rika subject matter. This group of courses meets for two 50-minute classes a week for 35 weeks. The second group of courses is more academically oriented. These courses are mainly for the students wanting to go to national colleges and universities, whose entrance examination include Rika subjects; or for students wanting to attend science or science related faculties at private colleges and universities where, of course, the entrance examinations include Rika subjects. Each of these courses meets for four 50-minute classes a week for 35 weeks. The third group is an advanced course set for students who will enroll in the university science and technology faculties. Each of these courses meets for two 50-minute classes a week for 35 weeks. The last category has only one course, Sogo-Rika or Integrated Science, and its aim is to cultivate a total scientific view of nature. It is a Monbusho approved activity-oriented course but still the number of upper secondary schools offering the course is quite small. There are several reasons. First, there are no textbooks authorized by Monbusho for the course. Second, there are few science teachers with the preparation needed to teach the course. Third, the course is not represented in the entrance examinations for most colleges and universities.

The first impression this description gives is that Rika, quite simply, is the Japanese equivalent of "school science" in any Western country, and indeed that is what most Japanese science educators believe about Rika. The similarities, however, are only superficial. According to the 1989 Gakushu Shido Yoryo (course of study) for elementary schools,

> Rika encourages pupils to: [1] *commune* with Shizen (nature), [2] perform observations and experiments, [3] acquire the ability of problem-solving, [4] acquire the *feeling of loving* Shizen (nature), [5] understand natural things and phenomena, and [6] acquire the scientific view and way of thinking. (Tr. by author, emphasis added)

These six objectives can be categorized into two groups, (2), (3), (5), and (6), and (1) and (4). The objectives in the first group are very similar to any objectives relating to science education. The objectives in the second group are Shizen *education* objectives. Thus, Rika is a melding of different traditions (which I will amplify in the following section) that results in a distinctively Japanese form of science education.

"Neo-science" Education and Shizen Education

The first group of objectives appears to be closely linked with the Western spirit of science education. Even here, however, important cultural realities influence the meaning and implementation of what seems so similar to Western science education. Take as an example objective "[2] perform observations and experiments." Under this objective, similar to what is done in Western science education, Japanese teachers have students do scientific observation, scientific experiment, scientific problem-solving, for the purpose of gaining a scientific understanding of natural things and phenomena. In contrast to Western science education, Japanese teachers practice what I will call, "neo-science" education. Neo-science means that teachers use instructional activities that the teachers think are models of science as science is understood in the West – but they are not. Pupils are not simply asked *to observe* as in "to be or become aware of, especially through careful and directed attention; notice."[88] In "neo-science" education, pupils perform and *enjoy* activities such as "observation" and "experiment" but the spirit of Western science does not guide them. Pupils are guided by Japanese motives. The teachers nevertheless view the activities as being done in the spirit of Western laboratory science. For the Japanese students and their teachers, *enjoying* activities in and of themselves is a principle and appropriate aim. Pupils have only to enjoy such activities. The activities do not necessarily need to culminate in theoretical abstractions nor to affirm any hypotheses. The Japanese priority is for pupils to have joy in Shizen and in the activities themselves. If this is accomplished, the lesson is not lost if some pupils never acquire the theoretical abstractions of science or the scientific facts.

Following the pattern set by Muraoka and Takamine, Japanese science teachers typically think of these Japanese objectives and activities as in the spirit of Western science. The difficulty is that the tradition of Kansatsu emphasizes the emotional and thus students are not "culturally" prepared to draw cognitive conclusions from their observations. In short, for the Japanese, the purpose of activities such as observing or doing experiments is the activity itself. The student tends not to think of the activity as the means for other purposes. School experiences with "scientific experiment" easily turn out to be a rather simple trial or experience, for example, without any intention that there be a discovering or a finding out of something new, or testing the truth or falsity of an idea. Indeed, "Jikken", the Japanese translation of "scientific experiment" has several meanings. The first one is to try to do something in practice, the second is to see or hear about the fact or reality of something, and only the last one – and least historical meaning – is scientific experiment. As one would expect, the students are more comfortable with the more historical understanding of Jikken.

Like Neo-science, the Japanese tend to think that their own concept of Shizen as a properly scientific concept. As I have stated already, it is for most Japanese the equivalent to the Western concept of "nature." In Rika objectives 1 and 4 one has Shizen education that means that pupils are to learn by direct interaction with Shizen, feel Shizen, feel empathy with Shizen, and to love Shizen. It is in the study of living organisms that Shizen education in Rika is most visible. For example, third and fourth graders cultivate plants such as sunflower, loofah, and Japanese morning glory. They plant and water seeds. They keep watch and take notes on how the seedlings are growing. They draw pictures of their plants. *They even talk to the plants*. They come to think of the plants like family members. They are encouraged to show love for their plants through the care they give to the plants. That is what is expected. When asked why do plants grow better in sunny weather than in cloudy or

[88] *The American Heritage Dictionary of the English Language*, Third Edition is licensed from Houghton Mifflin Company. Copyright © 1992 by Houghton Mifflin Company.

rainy weather? Pupils frequently answer: "Because they love sunny days just as we do." Or, "Because they need warmth to become vital." Teachers respect such answers. Third graders keep and care for butterflies in the classroom. They take cabbage leaves with butterfly eggs and watch them hatch. They take care of the larva by giving fresh cabbage leaves. They watch the process of pupating, emergence, and at last, they allow the butterfly to fly back into the free sky. They watch the whole life process of the butterfly. They report the color, size, and form of the larva. However, what they act out is not "scientific observation" in the Western sense but their love for the butterfly. In some cases children even name the butterflies.[89] Teachers promote such attitudes and activities among the pupils. The fact that these are not scientific observations is first of all not recognized. The teachers think of what they are doing as science education. Second, it is irrelevant to the cultural practices of Shizen education where the aim is to commune with Shizen.

Western science educators may find it curious that Shizen education is part of Rika since this view of nature is so foreign to the school science objectives in Western countries. *Rika, however, is not Western science education.* Rika is Japanese education that helps to transmit the historical Japanese sense of Shizen and Japanese sense of what the West calls nature, and lays the foundation for a *distinctively Japanese view of science.* The rationale for Rika and Shizen education lies deep within Japanese history and culture and what one sees in Japanese education today[90] is largely a product of the distinctively Japanese interpretations that Muraoka Han'ichi and Takamine Hideo had for their experiences with Western science.

The pattern of interest that pupils take in Rika is worth noting. Rika at the elementary grades shows more of the influence of Japanese culture, i.e., Japanese neo-science education and Shizen education predominate. Rika at the secondary level is much more like Western science education, in other words, the amount of neo-science education and Shizen education in Rika decreases. With respect to pupil interest, most pupils love Rika at the early elementary levels. As they move toward secondary levels, pupil interest declines (Matsubara et al., 1992); as I have argued elsewhere, science education as defined in Western terms is a foreign knowledge system for the Japanese (Ogawa, 1995b).

Shizen: The Japanese Worldview

Until in the 1970s, the landscapes of rural Japan was largely unchanged from the time of the Tokugawa era (17th-19th centuries), which in effect means that rural Japan had not changed all that much from the Jomon and Yayoi eras. In the 1970s, Japan came into the age of modern economic development. With this development came the large-scale destruction of Japan's natural environment. Today issues of pollution

---

[89] Readers, who remember similar science activities in their own culture, should be reminded that such scenes emerge from their own cultural backgrounds, not from the Japanese background of Shizen. A similar action can occur in two different cultures. The critical questions are: Is the meaning of the action the same in both cultures? Are the reasons for the action the same in both cultures?

[90] There are other non-western, science educators who will feel a kind of sympathy with Shizen education. Korea, for example, has similar aims and objectives for school science (Ogawa 1986b). However, none of the western documents referring to the aims, objectives, and goals of school science such as Project 2061, or NSTA-SSC or the new *National Science Education Standards* in the USA have anything even remotely similar to the Japanese notion of Shizen as a rationale for science education.

are of critical importance in Japan and the Japanese intelligentsia has begun to ask what Shizen means for Japanese.

Shizen is Japan's "collective representation" (Ogawa, 1995b, p. 588) of what the world is really like and how one should relate to that world. Minamoto (1985) writes that for the Japanese Shizen is not only a beautiful landscape but also the root of life and even has the characteristics of religion. Shizen is the Japanese cosmology. In this sense, Shizen is neither the collocation of natural things nor phenomena, each of which can be the object of Western scientific investigation. Nor is it the sum of all linkages among objects. The Japanese anthropologist, Iwata (1989) argues that among the Japanese, as well as most Southeast Asian peoples, there exists a feeling of animism, even when the people are Buddhist or Christian. Everything surrounding human life, the mountains or hills, rivers, earth, plants, trees, insects, fish, or animals, has its own spirit (kami in Japanese), with which people can be communicate. Most Japanese feel and are familiar with such spirits. Again, with this kind of feeling and familiarity one cannot regard natural things merely as the objects of Western modern science.

What others would call school science education, in Japan *needs to remain* Shizen education. Because Shizen is the traditional sense of the interrelationship between people and natural environments, science needs to be interpreted within the context of Shizen. To the extent the Western science resists such interpretations it should be taught as a type of instrumental or practical knowledge system. This idea is similar to the Western concept of nature study popular in the late 19th century and early 20th century (Bailey, 1911; Bainbridge, 1978; Jenkins, 1981). The fact that Western societies abandoned this approach is no reason for Japan and other Asian countries to do likewise. Teaching science through Rika classes which are infused with the Japanese tradition of Shizen will help the Japanese keep their indigenous identity - an identity that will help to prevent the dangerous ecological habits Japan has of late fallen prey to - and concurrently learn modern science (Ogawa, 1986a & 1995b). I have argued elsewhere that this is a *multi*-science perspective (Ogawa, 1995b).[91]

Concluding Comments

As I have now come to the end of this long discussion of science education in Japanese culture, I emphasize again the main point of this story. The ideas and methods of modern science as developed in the West can be adequately taught in Japanese schools within the traditional Japanese worldview of Shizen – even though the result, Rika classes, may appear strikingly different from the Western practice of science education. Moreover, important Japanese concerns such as environmental protection and maintenance of the indigenous Japanese identity demand that science be understood in Japan from the perspective of Shizen.

In the past such statements would have been considered scientific heresy. The history of the global spread of modern science and science education is the history of efforts to rid education of indigenous factors considered irrelevant, if not a

[91] In other words, there can be many formulations of science and science education depending on cultural context. For this purpose, a broader definition of science is useful, that is, a rational perceiving of reality where "rational" is neither exhausted by nor limited to Western modern rationality.

hindrance, to Western modern science. Science educators had no doubt that science was above culture. The influence of science studies and constructivism has brought about the reexamination these received notions and brought new awareness of the importance of culture. As I have tried to demonstrate in my "Epic description" culture is something each group of people should capitalize upon rather than shun as outmoded in this modern age. And this process begins with a simple, yet profound question, *What should science education be like for us?*

REFERENCES

Ahagon, C. (1989). Wagakuni no Meiji Shoki ni okeru Pestalozzi Shugi Kyoiku Juyo ni kansuru ichi Kosatsu: Takamine Hideo no Jitsubutsu Kyojuron no ichi Haikei wo Chushin to shite. (Study of Acceptance of Pestalozzian Education in Early Meiji Period: On the Background of Takamine's Theory of the Object Lessons.) *Ryukyu Daigaku Hobungakubu Kiyo, Shakaigaku Hen* (Bulletin of Ryukyu University, Faculty of Laws and Literature, Sociology Section, *31*, 97-131. (In Japanese)

Agassiz, E. C. (1886). Louis Agassiz: His Life and Correspondence. 2vols., Boston.

Anzai, T. (1995). *Fukuzawa Yukichi to Seio Shiso* (Fukuzawa Yukichi and Western Thoughts). Nagoya Daigaku Shuppankai: Nagoya, Japan. (In Japanese)

Bailey, L.H. (1911). *The nature study idea: An interpretation of the new school movement to put the child in sympathy with nature*. (4th edition, originally published in 1903) Norwood Press: Norwood, Massachusetts.

Bainbridge, J.W. (1978). Origin of the nature study movement. *Natural Science in Schools, 16*, 11-14.

Cipolla, C. M. (1969). *Literacy and development in the west*. Baltimore, MD: Penguin Books.

Cobern, W. W. (1991). *World view theory and science education research*. (NARST monograph No.3), National Association for Research in Science Teaching, Manhattan, KS: National Association for Research in Science Teaching.

Elkana, Y. (1971). The Problem of Knowledge. *Studium Generale, 24*, 1426-1439.

Elkana, Y. (1981). A programmatic attempt at an anthropology of knowledge. In Mendelsohn, E. & Elkana, Y. (eds.) *Science and Cultures: Anthropological and Historical Studies of the Sciences*. D. Reidel Publishing Company: Dordrecht, Holland, 1-76.

Hasegawa, M. (1986). *Karagokoro* (Mind of Chinese). Chuo Koron Sha: Tokyo, Japan. (In Japanese).

Horton, R. (1967). African traditional thought and Western science. *Africa, 37*, 50-71, and 155-187.

Hodson, D. (1993). In search of a rationale for multicultural science education. *Science Education, 77*(6), 685-711.

Irimoto, T. (1996). *Bunka no* Shizenshi (Natural History of Culture). Tokyo Daigaku Shuppankai: Tokyo. (In Japanese)

Ishikawa, H. (1902). Joshi Koto Shihan Gakko Kocho Takamine Hideo Kun (The Principal of the Woman's Senior Normal School, Mr Takamine Hideo). *Kyoiku Kai, 1*(11), 73. (In Japanese)

Itakura, K. (1968). *Nihon* Rika *Kyoiku Shi* (History of Japanese science education). Daiichi Hoki Shuppan: Tokyo, Japan. (In Japanese).

Iwasaki, T. (1977). Doitsu Teikoku to Kyoiku no Kindaika (German Empire and Modernization of Education). In Uneme S. (ed.) *Doitsu Kyoikushi 2* (History of Education in Germany, Vol.2), 33-75, Kodansha: Tokyo. (In Japanese)

Iwata, K. (1989). *Kami to Kami: Animizumu Uchu no Tabi* (Spirits and Gods: Travel in the Universe of Animism). Kodansha: Tokyo, Japan. (In Japanese).

Iyama, H. (1989). Kagakusha no Tanjo to sono Hohoteki Jikaku (Birth of Scientist and Methodological Awareness). In Ito, S. and Murakami, Y. (eds.) *Seio Kagakushi no Iso* (Perspectives of History of Western Science), 353-373, Baifukan: Tokyo, Japan. (In Japanese).

Jenkins, E.W. (1981). Science, sentimentalism or social control?: The nature study movement in England and Wales, 1899-1914. *History of Education, 10*, 33-43.

Junge, F. (1885). *Der Dorfteich als Lebensgemeinschaft*. Lipsius und Tischer: Keele, Prussia. (Cited from Itakura (1968), 190-199.)

Kaigo M. (ed.) (1971). *Nihon Kindai Kyoikushi Jiten* (Encycropedia of the History of Modern Japanese Education). Heibon Sha: Tokyo, Japan (In Japanese).

Kawasaki, K. (1990). A Hidden conflict between Western and traditional concepts of Nature in science education in Japan. *Bulletin of School of Education, Okayama University, 83*, 203-214.

Kawasaki, K. (1992). Kansatsu no Kenkyu (An epistemological study on "Kansatsu" believed to be a precise equivalent for "Observation". *Nihon* Rika *Kyoiku Gakkai Kenkyu Kiyo* (Bulletin of Society of Japanese Science Teaching), *33*, 71-80. (In Japanese).

Kitoh, S. (1996). Shizen *Hogo wo Toinaosu* (Reconsidering Natural Protection). Chikuma Shobo: Tokyo, Japan. (In Japanese).

Kobayashi, T. (1976). *Society, schools, and progress in Japan*. Pergamon Press: NY.

Kosakai, T. (1996). *Ibunka Juyo no Paradox* (Paradox in Acceptance of Foreign Culture). Asahi Shinbunsha: Tokyo, Japan. (In Japanese).

Lovejoy, A.O. (1936). *The great chain of being: A study of the history of an idea*. Harvard University Press: Cambridge, MA.

Matsubara, S., Kakisawa, M., Masuyama, H. and Ogiwara, M. (1992). Rika ni kansuru Kyomi-Kanshin to Seiseki tono Kankei (Relations between students' interests and achievements in Rika.) In Umeno, K. (ed.) *Longitudinal Survey Research of Students' Qualititative Changes of Scientific Attitudes and Science Learning*. National Institute for Educational Research: Tokyo, Japan, 26-31. (In Japanese).

Minamoto, R. (1985). Nihonjin no Shizen-kan (Japanese View of Shizen). in Shizen *to Kosmosu* (Shin-Iwanami Koza Tetsugaku No.5) (Shizen and Cosmos: New Series of Iwanami Philosophy, No.5), 348-374, Iwanami Shoten: Tokyo, Japan (In Japanese).

Murakami, Y. (1994). *Kagakusha towa Nanika* (What is Scientist?). Shinchosha: Tokyo, Japan. (In Japanese).

Muraoka, H. (1880). *Heimin Gakko Ronryaku*. (Japanese Translation version of *Die Praxis der Volksschule* by C. Kehr). Monbusho: Tokyo. (In Japanese).

Muraoka, H. (1883). Butsurigaku Kyoiku Ho (Teaching method of physics). *Toyo Gakugei Zasshi*, (In Japanese). (Cited From Itakura (1968), 153-156.)

Murayama, H. (1978). *Oswego Undo no Kenkyu* (A Study of the Oswego Movement). Kazama Shobo: Tokyo, Japan. (In Japanese with English Summary)

Murthy, N. (1973). *The Rise of Modern nationalism in Japan: A Historical Study of the Role of Education in the Making of Modern Japan*. Ashajanak Publications: New Delhi, India.

Nakamura, H. (1964). *Ways of Thinking of Eastern Peoples: India-China-Tibet-Japan*. East-West Center Press: Honolulu, Hawaii.

Ogawa, M. (1986a). Toward a new rationale of science education in a non-Western society. *European Journal of Science Education*, *8*, 113-119.

Ogawa, M. (1986b). Rika Kyoiku niokeru Kagi Gainen, "Shizen" wo megutte (A preliminary study on a key concept, "Shizen" involved in science education). *Ibaraki Daigaku Kyoiku Gakubu Kiyo* (Bulletin of Faculty of Education, Ibaraki University; Educational Sciences Devision), *35*, 1-8. (In Japanese).

Ogawa, M. (1995a). Kagaku wo Manabu Kachi wo Megutte (What is the value or worth of learning science?: From the perspective of "science as the culture of scientific community"), *Kagaku Kyoiku Kenkyu* (Journal of Science Education in Japan), *19*(1),19-27. (In Japanese).

Ogawa, M. (1995b). Science education in a multiscience perspective. *Science Education*, *79*, 583-593.

Oishi, S. & Nakane, C. (1986). *Edo Jidai to Kindaika* (Tokugawa period and modernization). Chikuma Shobo: Tokyo, Japan. (In Japanese).

Okamoto, M. & Mori, I. (1976). Rika Kyoiku ni Arawareta Wagakuni no Dentoteki Shizenkan (Traditional view of Shizen appearing in science education in Japan), *Kagakushi Kenkyu* (Journal of History of Science, Japan), *118*, 98-101. (In Japanese).

Sasaki, T. (1988). Nihon ni okeru Hatasaku Noko no Seiritsu wo megutte (On the Establishment of Field Agriculture in Japan). In Sasaki, T., & Matsuyama, T. (eds.). *Hatasaku Bunka no Tanjo* (New Born of the Culture of Field Agriculture), 1-22, Nihon Hoso Shuppan Kyokai: Tokyo. (In Japanese)

Sasaki, T. (1991). *Nihonshi Tanjo* (Dawn of Japanese History). Shueisha: Tokyo, Japan. (In Japanese)

Sasaki, T., & Matsuyama, T. (eds.). (1988). *Hatasaku Bunka no Tanjo* (New Born of the Culture of Field Agriculture), 1-22, Nihon Hoso Shuppan Kyokai: Tokyo. (In Japanese)

Shimada, Y. (1993). *Ijigenkokan no Seiji-jinruigaku:Jinruigakuteki Shiko towa Nanika* (Political Anthropology of Hetero-dimensional Exchange: What is Anthropological thinking?). Keiso Shobo: Tokyo. (In Japanese).

Tada, K. (1992). *Manabiya no Tanjo* (Establishment of Schools). Tamagawa Daigaku Shuppanbu: Tokyo, Japan. (In Japanese).

Turner, D. M. (1927). *History of Science Teaching in England*. Chapman & Hall Ltd: London, UK.

Watanabe, M. (1992). *Kagaku no Ayumi, Kagaku tono Deai (ge)* (Progress of Science, Encounter with Science (part 2)), Baihukan: Tokyo, Japan. (In Japanese).
Wilson, B. (1981). *Cultural contexts of science and mathematics education: A bibliographic guide*. Centre for Studies in Science Education, University of Leeds:Leeds, UK.
Yanabu A. (1982). *Honyakugo Seiritsu Jijo* (How did translation words establish?), Iwanami Shoten: Tokyo, Japan. (In Japanese).
Yasuda, Y. (1993). Retto no Shizen Kankyo (Natural Environments of Japanese Islands). In Iwanami Koza: Nihon Tsushi (Iwanami Series: History of Japan): Vol.1. *Nihon Retto to Jinrui Shakai* (Japan Islands and Human Society), 41-81, Iwanami Shoten: Tokyo, Japan. (In Japanese).

Gürol Irzik

Chapter 8

Philosophy of Science and Radical Intellectual Islam in Turkey[92]

Every culture, every nation can build a science that fits its own particular needs
Paul Feyerabend (1991)

1. Introduction

Teaching philosophy of science is not merely a self-subsistent epistemological activity, but at the same time a political one. At least, this is what I am going to argue. As I see it, the intertwining of epistemological and political dimensions in philosophy of science is contingent, not logical. That is, I believe that while it is not possible to derive political views from epistemological assumptions logically, it is nevertheless often the case that the two get allied in interesting ways in different contexts. This is particularly conspicuous in the case of Turkey due to historical reasons and is closely related to Islamists' changing attitudes toward the process of Westernization. In section 2, I briefly describe the historical context of Westernization from the mid-nineteenth century Ottoman Empire to the formation of Turkish Republic.

Roughly speaking, we can delineate two distinct positions or attitudes toward the process of modernization among the Islamists. According to the first attitude, there is no conflict between Islam and modern industry, science and technology as the latter are taken to be the motor of economic development. This attitude has prevailed for more than a century and is still the dominant one. However, in the last fifteen years, a new attitude emerged in the writings of a number of Moslem intellectuals: Islam turned into a powerful critique of and became irreconcilable with all aspects of modernization, including "Western" science and technology. I discuss these opposing attitudes in sections 3 and 4, respectively.

The newly emerging position rests on a wholesale rejection of modernity and draws on diverse sources, most notably, post-positivist and postmodernist philosophy and thought. Section 5 is devoted to identifying these sources. The critique of existing ("Western") science and technology as part and parcel of modernity naturally raises the question of an alternative conception of science and technology

[92] Part of the research for this paper was done at the Center for Philosophy of Science, Pittsburgh University. I am grateful to its director Gerald Massey for his invitation and continuous support. I also acknowledge the financial support of Boğaziçi University Foundation. Bill Cobern's encouragement was essential for this project. My warmest thanks to him.

W. W. Cobern (ed.), Socio-Cultural Perspectives on Science Education, 163–179.

Section 6 is a discussion of the possibility of an Islamic science based on Islamic values.

The Islamists' critique of modern science and technology resonate forcefully among students, and consequently the classroom turns into a lively forum in which different conceptions of science and technology, different philosophies of science clash against each other and get allied with different worldviews. I discuss this phenomenon in section 7. Finally, I conclude by offering some suggestions about what role philosophy can play within such a context.

2. Historical Background

The Ottoman Empire was in serious military and economic crisis in the nineteenth century. On the one hand, it was losing big chunks of territory; on the other hand, its pre-capitalist economy was utterly failing to compete with the industrialized Western countries. Indeed, the very existence of the pre-modern Ottoman Empire was threatened by the modernized, technologically and economically developed capitalist nation-states in Europe.

Accepting the superiority of the West and perceiving the industrial, technological and scientific development behind its power, the Ottoman ruling elite initiated a series of reforms which aimed to modernize (or, which amounts to the same thing, Westernize) the state and the country. Not surprisingly, the reforms received mixed reactions from the intellectuals and bureaucrats. Some, the Westernists, as they are called, looking to the Western lifestyle with admiration, supported them in all spheres. Others, notably Islamists who constituted the majority, while endorsing the reforms in the industrial, technological and scientific domains, rejected their extension to the political, cultural and moral life which was based on Islamic values and norms. As Islam constituted the political identity and the legitimizing ideology of the Ottoman Empire, they saw such an extension of reforms as endangering both their identity and the power structure of the State. They argued that Ottomans had nothing to borrow from the degenerate culture of the West, its materialistic and individualistic values that were in direct opposition to Islamic ones.

The Islamists' attitude towards the Western civilization can be formulated in terms of the form-content distinction: Science, technology and industry constituted the form; the political, social and moral values and practices formed the content. Because they believed that the form was neutral and separable from the content, they saw no difficulty in accepting the form and rejecting the content. They thought that by such a split they could fight back against the Western powers with their own weapons and yet retain their identity.

This was the dominant view among the Islamists and prevailed for more than a century. For example, in the second half of the nineteenth century Namik Kemal, the eminent Ottoman poet, writer and intellectual of the period, tried to incorporate Islamic jurisprudential practices into the newly formed constitutional structure. He wrote the first Turkish novel, *Intibah*, with the explicit intention of using the novel (which was obviously a Western literary form) as a most convenient way of promoting traditional-Islamic social and cultural norms against the dangers of Western values. During the first quarter of the twentieth century Ziya Gökalp, perhaps the most original Turkish sociologist and thinker who tried to synthesize nationalism, religion and modernism, distinguished between culture and civilization. According to him, while Islam was an integral part of the Turkish culture, the Turkish nation-state that he fought for belonged to the (Western) civilization that included science, technology and industry. Finally, the idea that Turkey needed these latter "universal practices" for the purposes of modernization but not the cultural and moral values of the West was the main theme of both National Salvation Party in the seventies and Motherland Party (which won the elections twice) in the eighties.[93]

At any rate, the struggle between the Westernists and the Islamists ended in favor of the former with the formation of the Turkish Republic in 1923 and implementation of the radical reforms in all spheres of Turkish society, under the leadership of Mustafa Kemal Atatürk. The modern nation-state replaced what was left of the Ottoman Empire, nationalism replaced Islam as the legitimizing ideology for political restructuring. As a result, a secularist nation-state, a state-fueled capitalist economy and a society on the basis of a new identity (Turkish as opposed to Islamist-Ottoman) slowly emerged.[94]

3. Early Reconciliation of Modernity with Islam

What is particularly relevant to our concerns is that the Westernist ruling elite blended a positivist ideology with the discourse of Enlightenment and modernism, according to which a rigorous scientific education could make it possible to control and transform both nature and society for the benefit of the people and to fight against superstitious beliefs, religious dogmas and traditions which stand in the way of modernization and social progress. Thus, positive (i.e., empirical) sciences became the only kind of activity which provides genuine knowledge about the world we live in.[95] Although the founders of the Republic never spelled out their positivist philosophy (after all, politicians are --pace Plato--rarely philosophers), the following views are implicit in their writings and speeches. Science is based on neutral observations and produces universal, objective and secure knowledge by its method. The world is a world of cold facts; it has an objective structure that transcends individuals and in no way depends on theories, languages, and practices. The aim of science is to describe this structure as accurately as possible, to discover its regularities in order to predict and control events in the world. Control is to be achieved by the technological applications of science, and science-driven technology in turn is expected to lead to a complete industrialization of the country, which is regarded as the key to becoming a prosperous modern (i.e., Western) society.

Three aspects of such a conception of science are important for our purposes. First, the source (whether divine or not) of theories in particular and beliefs in general is irrelevant. What matters is whether they are justified or not by observation and experimentation. This is the famous positivist distinction between the context of

---

[93] For Kemal's political views see Mardin (1962). J. Parla (1990) presents a fascinating epistemological examination not only of Intibah but also other novels of the period. Gokalp's views can be found in T. Parla (1985) and Davison (1995). For the programs of National Salvation Party and Motherland Party see (Toprak, 1984, 1988; Göle, 1993).

[94] See Gülalp (1995a), Keyder (1987) and Mardin (1981).

[95] The following saying by Atatürk, which decorates almost every school wall even today, encapsulates this well: "The most genuine guide (to truth) in life is science."

discovery and the context of justification. Second, a sharp division between theories, languages and practices on the one hand and the world of facts which is not in the least constituted by them on the other hand results in an image of science as a disinterested contemplation of an already structured, ready-made world. This means that science is a purely epistemological activity which ideally does not involve the race, religion, ideology, gender, personal or social interests of the scientist. Third, a similar dichotomy is held between facts and values. Accordingly, it is impossible to derive values and value judgments from facts. It is the business of science to supply the facts about the world, but not the values and value judgments.

Now, how is this positivistic understanding to be reconciled with Islam? From the traditional, pre-1980 Islamists' point of view, the reconciliation is made possible largely on the basis of the fact-value distinction. According to this point of view, while science gives the value-free facts, Islam supplies the values themselves, especially the moral ones. It is because of their belief in this division of labor between science and religion that the traditional Islamists wholeheartedly welcomed the scientific, technological and the industrial developments of the West and rejected the Western moral and cultural values. The Westernists, on the other hand, reconciled the positivistic ideology with Islam by separating knowledge from faith, the this-worldliness of science from the other-worldliness of religion, and at the same time turning religion, as much as possible, into a private affair between the believer and the believed. Needless to say, the Westernist Republicans of the 1920's and 30's did not abolish the system of *Imam*-hood as a mediator, nor did they prohibit people from attending mosques; but they exercised a strong control by making Imams civil servants who were to follow the official interpretation of Islam and various practical directives of the Higher Institute of Islam which was also a state institution.

So, it was argued as an semi-official state policy that science gives us knowledge about the world in which we live, that religion governs the realm of faith and prepares us for another world, another life after death. Islamic views concerning the creation of the universe, the origin of human and other species, for example, were interpreted not literally but metaphorically so that they did not conflict with the scientific theories in astrophysics and evolutionary biology. Religion and science were taken as different languages and discourses functioning at different levels, each legitimate and valid in its own domain. In such a context, there is no substantial difference between teaching philosophy of science in Turkey and, say, in France or the USA.

4. The Radical Intellectual Islam: Critique of Modernity

The situation, however, began to change dramatically after the 1980's when a number of well-educated, young Moslem intellectuals (most notably, Ali Bulaç, Ilhan Kutluer, Rasim Özdenören, and Ismet Özel) rejected this kind of a marriage between science and religion. Radical Intellectual Islamism as I shall call their views for the lack of a better term, is as much a critique of Western industry, technology and science as it is of the old Islamists' attitude towards Westernization outlined above.[96] This is because, according to Radical Intellectual Islamists, modernity and science are inextricably linked. For example, Ilhan Kutluer (1985, p. 9) writes that:

> Today, the terms "modern" and "scientific" are often used synonymously. For that reason, "modern age" is taken to mean "scientific age". By this is meant both that a new conception has started a new age and that scientific practices have left their stamp on a new age.

Indeed, as we shall see below, Radical Intellectual Islamism is a powerful but an eclectic critique of modernity in general, which draws on diverse sources ranging from a certain interpretation of Islam to postmodernism.

The young Moslem intellectuals criticize their predecessors on the grounds that they failed to see the unity of form and content, that is, the inseparability of Western industry, technology and science from the broader cultural and intellectual values (such as materialism, Cartesian dualism, individualism and secularism) which nourish them. From this perspective it is simply naive to embrace the former while rejecting the latter (Özel 1992, p. 45-47; Bulaç, 1995a, p. 320).

These intellectuals also criticize industrialization on the grounds that production for the sake of ever more profits has turned human beings into mere puppets of the Capitalist consumer society manipulated by mass media, deprived them of their religious/spiritual values, and enslaved them to the greed for material wealth. Science and technology have also received their share from this sweeping criticism since they are seen as the driving motors behind the ever increasing rationalization of all aspects of social and economic life. Positivistic science and technology, the Islamist intellectuals argue, secularized social life, but utterly failed to establish any moral order; with the age of Enlightenment science itself became a new religion but one without an ethics.[97] Westernist modernity, they claim, is destroying not only people's traditional customs, lifestyles and characters, but also the environment. As Ali Bulaç, one of the most distinguished and prolific Moslem intellectuals, boldly puts it: "Modernism, which promised a 'Paradise on Earth' turned the entire planet into hell.... It is impossible to unify Religion and Modernism over a common denominator" (1995a, p. 7-8).[98]

Radical Intellectual Islamists situate science and technology as crucial elements within the narrative of modernity; modernity provides both the content and the context for a certain conception of science and technology, which in turn serve to perpetuate modernity. A critique of the former, therefore, is ipso facto a critique of the latter. Conversely, we can say that Radical Intellectual Islamists' anti-science and anti-technology discourse is directed, above all, towards the modernist "ideology", the modernist system of values behind modern science and technology. What exactly is this "ideological matrix" from the viewpoint of Islamist intellectuals? As far as I can see, its elements are a dualist, mechanist and reductionist philosophy of being; constitution of the human mind (reason, in particular) as both the source and foundation of knowledge and morality (i.e., secularism); rationalism, progressivism, materialism and indivi-dualism.

---

[96] For a detailed summary of the various views of these intellectuals and their interesting educational backgrounds see Toprak (1993).

[97] See Bulaç (1995a, 1995b), Özdenören (1995) and Özel (1992).

[98] Note that Bulaç uses the term "modernism" in the sense of "modernity". This is typical of Islamist writers.

(a) Dualism and mechanism: As is well known, the rise of modern science in the sixteenth and seventeenth centuries is based on a new conception of being, namely, Cartesian dualism and mechanism. Cartesian dualism destroys the unity of being by declaring matter and mind as two distinct and independent substances. While the essential property of matter is extension, that of mind is thinking. Since only human beings think, everything else – the world, planets, stars, and even the human body – is just extended matter. The entire universe is nothing but a huge machine subject to mechanical laws. According to Islamists, it is precisely such a conception of nature devoid of any divine meaning and order which legitimizes the scientific-technological control, domination and exploitation of nature and thus paves the way for ecological destruction (Kutluer 1985, p. 70-75; Bulaç 1995a, p. 18-21).

(b) Reductionism: Modern science is also reductionist in the sense that it aims to under-stand the structure and function of the whole in terms of the structure and function of its parts. Reductionism stems from the Cartesian philosophy according to which all physical phenomena can be explained by the size, shape and motion of little particles out of which material objects are constituted. Reductionism gives support to the view that only that which can be measured or quantified can be explained and understood (Bulaç 1995a, p. 15).

(c) Worship of human reason, humanism and secularism: According to modern thought, the only source and foundation of knowledge is human reason broadly construed. Thus, although modern rationalists and empiricists are divided over the question of whether it is sense experience or innate concepts and principles that provide the ultimate source and ground of knowledge, they agree that it cannot be anything other than the faculties of human mind. Reason replaces revelation, faith, custom and habit; and science, as the embodiment of reason par excellence, becomes the only activity capable of producing knowledge about nature, society, and the individual (Özdenören 1995, p. 139-140; Özel 1992, p. 150). Similarly, human mind and reason also become the sole source and ground of rights, moral values and principles. Religion no longer regulates the domain of law and moral conduct. Man becomes the measure and the ultimate end of man (Özel 1992, p. 80).

(d) Instrumentalization of reason and rationalization: But modern reason is an instrumental reason. Rationality of human actions is a means-ends rationality. Any action is rational so long as it achieves a pre-given goal in the most efficient manner possible (Bulaç 1995a, p. 21-22, 314). Instrumentalization of reason goes along with a rationalization of the domains of economy, state institutions (bureaucratization), and public and individual life. Efficiency becomes the regulating principle of virtually everything, and accordingly, social sciences (especially, economics, sociology, and social psychology), which are the products of modernity, tend to be mere instruments for the efficient and productive organization of various social spheres and for rendering them more calculable.

(e) Progressivism: The idea of social, economic, industrial, technological and scientific progress – all to the benefit of humanity – is one of the most important aspects of modernity (Bulaç 1995a, p. 31-32; Kutluer 1985, p. 28-29; Özel 1992, p. 153-159). The narrative of progress as the emancipation of peoples serves conveniently to legitimate modernity as the latter is seen as an inevitable stage in the historical development of mankind. The idea of scientific-technological progress reinforces this narrative because, according to modernists, progress in other domains is made possible by the advances in science and technology; indeed, without them modern consumer society would simply stop existing.

(f) Individualism and materialism: The modern individual is an atomized, materialist, selfish, greedy, and alienated individual (Özdenören 1995, p. 10-19; Özel, 1992, p. 68-81). He has lost his unity and "spiritual, intellectual, metaphysical and cosmic reality" (Bulaç 1995a, p. 27). For that reason, the modern individual's humanity is an impoverished one. According to Islamist intellectuals, it is no coincidence that while secular science and technology become more and more mechanistic, reductionist, and instrumental, man becomes less and less human. For, modern science and technology have reduced man to a functional body that consists of a measurable, calculable swarm of atoms and genes.

(g) Finally, modern societies, Islamist intellectuals observe, are capitalist societies. Capitalism both nourishes and is at the same time nourished by secularism, instrumental reason, rationalization, materialism and individualism. To the extent to which Western science and technology serve these values, they serve capitalism (Bulaç, 1995b, p. 234-243).

This completes my documentation, in the Islamist intellectuals' writings, of the ideological matrix of modern science and technology against which their critique is directed. It is worth pointing out, however, that occasionally, Islamists' critique extends over the very products of technology, and the content of science, that is, scientific theories. For example, according to Ali Bulaç, people can and indeed have lived for centuries without TV's, refrigerators, computers, detergents, artificial food, clothes, fertilizers, plastic, cement and asphalt. These products are not at all necessary for survival; they are promoted by Western powers in order to encourage consumerism in traditional societies and thus make them dependent on Western economy and technology (Bulaç, 1995a, p. 39-42).

In a similar vein, Bulaç criticizes modernization theories in economics and sociology on the grounds that (a) they classify traditional societies as "backward", "primitive" and "underdeveloped" just because such societies are not industrialized and do not have sophisticated technologies, (b) they serve to justify Western industrial development as an inevitable stage which all countries must go through, and (c) they promote the dependence of traditional societies on Western economies and thereby legitimize Western imperialistic hegemony (Bulaç, 1995a, p. 84-89). Bulaç is also very critical of experimental psychology, psychoanalysis, and social psychology for objectifying the inner world of the individual, for their futile attempt to understand human psyche in terms of cause-effect relationships. His verdict is indeed harsh:

> It is impossible to objectify the emotional world. Since Wilhelm Wundt's experimentalization of psychology, since the movement which separated psychology from philosophy by reducing it to the methods of biology and physics, nothing serious and worthwhile has been said about the rich spiritual and emotional world of the individual (Bulaç, 1995a., p. 309-310).

5. Sources of the Critique of Modernity

As can be expected, this comprehensive critique of the ideological matrix of modernity does not originate merely from the resources of Islam (such as Qur'an and Sunnah, i.e., Prophet Mohammed's teachings); in fact, to a large extent it exploits a heterogeneous combination of non-Islamic writings to which I now turn.

One of these sources can be grouped under the label of Marxism: Marx, the Frankfurt School and Habermas. Radical Intellectual Islam draws on the Marxist critique of capitalist relations of property, exploitation of one class by another, consumerism, alienation, and fetishism. Bulaç writes, for example: "Fundamental relations of ownership in unjust and inhumane capitalism have not changed at all after Marx exposed the issue with all its nakedness" (Bulaç, 1995b, p. 47). As the sociologist Haldun Gülalp has rightly observed, "Özel links capitalism and consumerism with idolatry" (Gülalp, 1995b, p. 65) when he (Özel) writes:

> Now everywhere is a market. There is nothing which has not become a commodity that can be bought and sold. The market, with its impenetrable mechanism, its shrines, banks, production and consumption armies, serves as a god for those who have gone astray from religion. (Özel, 1992, p. 32)

Özel pays special attention to the problem of alienation. The subtitle of his book *Three Problems*, namely, "Techne-Civilization-Alienation" indicates the centrality of his concern with this issue. He discusses Hegel's, Feuerbach's, and Marx's views in this context at length and concludes that alienation is a consequence of Western humanism and atheism (Özel, 1992, p. 68-81).[99] Islamist intellectuals also rely heavily on the critiques of the negative aspects of modernity, especially the instrumental rationality, the Enlightenment idea of progress, scientism and positivism offered by the Adorno, Horkheimer, and Habermas (Bulaç, 1995a, p. 21, 206, 233). This does not mean, however, that Radical Intellectual Islamists are pro-Marxist. On the contrary, they see Marxism squarely within the discourse of modernity and thus denounce its humanism, secularism, atheism, materialism and its alternative program of socialism. In a similar vein, they find Habermas's attempt to remedy and complete the project of modernity futile.

The recent current of postmodernism provides another convenient source for Intellectual Islamism's critique of modernity. Writings of Lyotard, Baudrillard, Deleuze, Guattari, Derrida, and Illich are all cited approvingly. For example, Bulaç (1995a, p. 231-232) agrees with Lyotard that knowledge should not be identified with scientific knowledge and that scientific knowledge has become a commodity like everything else. He praises Lyotard for accurately describing the new (postmodern) nature of knowledge in the post-industrial societies in which knowledge and power have become one and for formulating the double problem of its legitimization: "Who decides what knowledge is, and who knows what needs to be decided?" (Lyotard 1984, p. 9). Following Ivan Illich, Bulaç also claims that the entire system of modern education from elementary schools to universities is an effective agent of modernity and constantly reproduces a modern lifestyle. He concurs with Deleuze and Guattari that the state has besieged the individual's private

[99] It may be worth noting that Ismet Özel was an eminent Marxist poet and writer until early seventies after which he converted to Islam and denounced Marxism. See his intellectual autobiography (1988).

life from all sides and penetrated deeply into the roots of his consciousness with the help of modern science. To resist this is to risk being schizophrenic (1995a, p. 116-121).

Although Islamist intellectuals successfully exploit postmodernism's total rejection of modernity, they are themselves no postmodernists:

> Now, postmodernism is against [social] determinism, rationalism and.... positivism and overlaps with us. But.... neither modernism nor postmodernism can be a cure.... Postmodernism detaches man and universe from their epistemological and moral content, from their Divine, sacred and transcendent goal. (Bulaç, 1995a, p. 238-239).

Postmodernism is, indeed, antithetical to Islam; for, as is well known, from the postmodern perspective the term "modern" designates,

> any science that legitimates itself with reference to a metadiscourse of this kind making an explicit appeal to some grand narrative, such as the dialectics of Spirit, the hermeneutics of meaning, the emancipation of the rational or working subject, or the creation of wealth. (Lyotard 1984, p. xxiii)

Therefore, modernity is seen as a unified, totalizing master narrative which suppresses difference, plurality, and dissension; accordingly, postmodern is defined as "incredulity towards metanarratives" (Lyotard 1984, p. xxiv). Islam, on the other hand, is nothing but an absolutist, totalizing narrative precisely in the sense intended by Lyotard and should therefore be met with equal skepticism. Nevertheless, it is not difficult to see why both Marxist and postmodern critiques of modernity appeal so much to Radical Islamist intellectuals: partly because such critiques come from within, and partly because they create a liberating effect. Modernity can no longer claim universal validity. This makes it possible to counterpose the notion of an Islamic civilization against the West and to present it as a new and promising alternative.

A third source of Radical Intellectual Islam is the post-positivist philosophy of science, in particular, Thomas Kuhn's and Paul Feyerabend's works. Islamist intellectuals widely and approvingly read them for obvious reasons. Both Kuhn and Feyerabend destroy the positivist picture of science. With his colorful metaphors like gestalt switches, religious conversions and puzzle solving, Kuhn undermines the standard rigid conceptions of scientific rationality, progress and truth, and narrows the gap between science and non-science, and a fortiori, between science and religion. If it is not possible to speak of a linear progress in science, it should be even more impossible to talk of a social progress as envisaged by modern thought (Kutluer, 1985, p. 174-180).

Similarly, Feyerabend's provocative attack on method and science resonates in Radical Intellectual Islam with full force. Feyerabend (1987, 1988) argues that there is no such thing as the universal scientific method if by "method" one means a set of fixed and supra-historical rules which are valid for and applicable to all kinds of problems, in every scientific context; every rule is context-bound and thus has its limits. Just as there is a plurality of "methods" in science, Feyerabend claims, (Western) science itself is just one tradition among many, one ideology among others. It is in no way superior to such practices as religion, magic and astrology. Feyerabend believes that since in the past numerous cultures have survived for thousands of years without the help of science, it is outrageous to impose science upon them at the expense of their traditions. In his preface to the Turkish translation of *Against Method*, comparing Moslem intellectuals with the proponents of the Japanese Enlightenment who adopted not only the practice of science but also its ideology, he says "Moslem radicals have a more reasonable attitude. They use the products of first world science but despise its ideology" (Feyerabend, 1989, p. 12). Feyerabend's pluralistic relativism finds its most succinct expression in his claim that "Every culture, every nation can build a science that fits its own particular needs" (Feyerabend, 1989, p. 11; see also 1991, p. 11-12). There are as many kinds of science as there are societies and no objective grounds for establishing the cognitive superiority of one over another.

Feyerabend's philosophy claims to unmask the ideology behind science by exposing its roots, i.e., the materialist Western civilization. Thereby it affirms the Islamists' conviction that science as it is commonly practiced today is a hegemonic Western tradition and paves the way for the conception of an alternative science based on Islamic values:

> Who could ignore the contributions anarchist methodologists made to philosophy of science and epistemology, courageously displaying modern science's despotic, reductionist, mechanist and monist real face? (Bulaç, 1995a, p. 142)

No wonder, then, that a Turkish translation of Feyerabend's *Science in a Free Society* was also published under the title "The Church of Science" (Bilim Kilisesi) and received enthusiastic reviews like "Scientific Despotism on Trial" in Islamist journals.[100]

Finally, Islam itself provides a rich source for the Intellectual Islamists, according to whom Islam is not only a guide for life in "the other world" but also in this world. They claim that Islam originated as a way of governing and structuring society in all spheres – intellectual, cultural, juridical, political and even economic; Islam is this-worldly through and through and cannot be divorced from its practical dimension. From this perspective Qur'an and Sunnah function not so much as source of a critique of Western modernity (though they do that, too) as the divine source of values and principles with which an alternative way of life, in particular, an alternative conception of science and technology can be envisioned.

6. Islamic Science Based on Islamic Values?

The task that faces the Islamist intellectuals then is, first, the articulation of a set of Islamic values as distinct from modern ones, second, of a conception of science and technology on the basis of these values, and, third, its implementation to develop a new form of science and technology. As they themselves admit, this is no easy task. But they believe that they have more or less completed the first stage. The Islamists' alternative conception has an ontological, an epistemological, and an ethical dimension and is based on the following values and principles.

The Ontological Dimension: God (Allah) is One, and he is the creator of everything. All being originates from a single arch, a single source, i.e., God. This is called the principle of unity (tawheed). Since the universe is the result of a divine creation, it is

[100] See Bilim Kilisesi-Özgür Bir Toplumda Bilim (Tr. by C. Cerit), Pınar Yayınevi, Istanbul (1991); M. Bilici, Scientific Despotism on Trial, Matbuat, 8-9, Dec. 1994-Jan. 1995, p. 35.

an organic whole which reflects God's unity. This means that every object, every event and process in the universe points to a divine Order and Reality. Thus, the principle of unity extends far beyond the unity of God and incorporates both the unity of soul and body and the unity of man and nature.

The Epistemological Dimension: The ultimate source of our knowledge of the universe is Qur'an, which consists of God's words, revealed directly to the Prophet Mohammed. Therefore, revelation is the highest form of knowledge. Scientific knowledge is not only inferior to revelation, but also subject to it in the sense that the aim of science is to reinforce man's belief in God and to help him become His servant by discovering the divine order. Understood in this way, the pursuit of knowledge (el-ilm) itself is a form of worship.

The Ethical Dimension: From these ontological and epistemological considerations follow man's epistemic and ethical humility. Because the universe is a divine creation, its mysteries transcend human reason. Our knowledge, including scientific knowledge, then, is bound to be limited. We should not naively believe that scientific and technological developments will solve all of our problems. Similarly, man should not disturb the order of the universe as this would be "arrogance". Accordingly, science and technology must be restrained so as not to destroy the natural environment; they must serve the individual and the community in harmony with the natural order of things.[101]

Would a new form of science and technology automatically arise out of these values and principles? What would it be like and in what ways, if any, would it differ from the current one? At the moment Islamist intellectuals do not have any answers to these questions, but they are becoming increasingly more sensitive to them. Kutluer, for example, cautions his readers against hasty solutions which come under the label of "Islamization of knowledge": "We cannot now reach a final decision concerning the necessary preconditions for an Islamic model of science, because we are not sure yet whether such a model is possible or not" (1985, p. 186). In a similar vein, Mustafa Armağan, who has recently translated and edited a collection of articles on Islamic science by non-Turkish Moslem authors, asserts in his introductory essay that the notion of Islamic science represents a movement against modernity, but then asks "will the Islamic wave be able to produce a genuine alternative?" (Armağan, 1990, p. 21).

These questions clearly indicate that Islamist intellectuals are well aware that the articulation of a set of Islamic values by itself does not mean much and that the program of Islamic science stands or falls with the actual construction of alternative science based on these values. Otherwise, the notion of "Islamic science" is bound to remain empty, and more importantly, the modern Western world is likely to continue its superiority over the Islamic one.

[101] For these ontological, epistemological, and ethical considerations see Bulaç (1995a, p. 311, 314, 322-326 and 1995b, p. 266-274, 306-309). A similar conception can also be found in non-Turkish Moslem authors; see, for example, Sardar (1984).

7. Teaching Philosophy of Science in the Classroom

Now I want to situate some of my teaching experiences against this background. This will not only give the reader an idea about the dissemination and reception among students of what I have called the Radical Intellectual Islam, but also enable me to show how any general stance concerning science involves political implications.

I have been teaching philosophy of science at Bogaziçi University in Istanbul since 1988. Our philosophy department requires all philosophy majors to take an introductory philosophy of science course in their fourth semester, but non-majors are allowed to take it as well. Most of these non-philosophy students come from the engineering school and the social science departments such as political science, sociology and psychology. During my first years of teaching, bringing my own experience from my graduate studies in the USA to the Turkish classroom, I used to begin with logical positivists, trace their development into a kind of critical empiricism, introduce Popperian falsificationism, and then discuss the post-positivist turn in the works of Kuhn, Lakatos, and Feyerabend. I would also spend some time on social studies of science and end the course by raising questions concerning the social, political and ethical dimensions of science.

This is more or less how I still teach the course, but the reactions I get from the students have begun to change. First of all, a few, though an increasing number of students argue that science has no privileged cognitive status and that it is on a par with, say, astrology or magic. Second, they question science itself, not just this or that interpretation of it. One of their favorite arguments against science is that technology based on it has brought our world near to its destruction by environmental pollution and nuclear arms. Third, they link modern science to the Western way of life, which they quickly blame for all the disasters in the world. Finally, they defend the Islamic way of life and claim that it provides the possibility of a better life and a more humane science.

The similarities between these arguments and conclusions on the one hand and those of the Islamist Intellectuals' on the other hand need no documentation. These students are obviously exploiting some of the ideas they have acquired in class and outside for their own purposes. As the language of instruction in our university is English, they have access to other relevant literature not read in class. But more importantly, since translations of not only Kuhn and Feyerabend but also Habermas, Foucault, Lyotard and Derrida have flooded the market and are now being discussed fervently among all intellectuals, and since these discussions take place within the context of debates about Turkish identity, democracy and modernity in general, the issues immediately capture the imagination of the students with a sense of reality and significance.

Indeed, it would be a mistake to think that purely intellectual motives are the only reasons that lead students and intellectuals alike to Radical Islam. With the liberalization of the economy, the switch from the import-substitution policy to the export-promotion policy, and the attempt to join the European Union and become integrated with the Western world, the modernization process has gained a new impetus in Turkey after the 1980 military intervention and brought about a number of economic, political and social problems which form the material basis of the rise of Radical Islam (see Gülalp, 1995a). More and more people are expressing dissatisfaction with these changes as inflation and unemployment rates soar, as the distribution of income gets worse and worse for the working people, and many think that the only way to overcome these problems is to return to our own roots, our own values, identity and religion.

In the last several years Turkish society is experiencing a dangerous tendency toward the polarization of its members into "laicists" and "Moslems". The polarization manifests itself in almost all spheres of life. For example, in the last ten years Moslem female college students stirred up a nationwide controversy when they began attending classes covered up in veils. Some experienced disciplinary action for violating the dress code passed by Atatürk in the late twenties. This gave rise to huge mass protests and subsequently counter protests by either side. But more tellingly, just three years ago thirty seven people, all "laicist" (which means "enemies of religion" from the "other" perspective) writers and poets were literally burnt to death when thousands provoked by Islamist militants put the hotel they were staying in on fire.

In such a context teaching philosophy of science necessarily involves a political dimension. It is political both in the broad sense and in the Foucauldian sense that it "acts upon the possibilities of action of other people" and "structures the possible field of action" (Foucault, 1983, p. 221). Adopting any systematic attitude towards science leaves out some possibilities and opens up others. For example, the positivistic conception cannot peacefully coexist with Radical Intellectual Islam; the latter excludes the former. Similarly, while a moderate understanding of Islam is largely neutral with respect to rival interpretations of science, Radical Islam is not. Consequently, teaching philosophy of science in an Islamic cultural context creates a burden of social responsibility in presenting, evaluating and criticizing alternative images of science.

8. Philosophy as Dialogue

How should one carry this burden? What is the best way of coping with a situation in which different conceptions of science are allied with radically different worldviews and ideologies? What role can philosophy of science have when it becomes continuous with politics?[102]

First, as philosophers of science we can draw attention to the multiplicity and heterogeneity of scientific disciplines. At one end of the spectrum is physics whose objects are amenable to mathematical and experimental treatment. At the other extreme is history as an interpretive discipline which is much closer to philosophy and literary criticism than to physics. In between lie various branches of science ranging from chemistry to biology, economics, sociology, psychology, anthropology and linguistics. Every scientific discipline has its own objects, problems, methods and aims, there being only partial overlap among them. What justifies the label "science" is a family resemblance among these diverse disciplines, rather than a set of necessary and sufficient characteristics.

[102] Notice that these questions are pertinent not only for teaching philosophy of science, but also for science education to the extent to which the latter involves an orientation toward the status of science in modern society, the relationship between science and technology, science and industry, and science and values in general (Jenkins 1992).

The recognition of this basic, yet often forgotten fact may caution against unfounded and sweeping critiques of science. For example, the Islamists' charge of mechanism may be valid for some of the physical sciences (even that is not quite right after the relativistic and quantum revolutions, though), but it is off the mark for history. Similarly, chemistry may be reductionist, but certainly not linguistics. It is hard to see how physics can be individualistic, though orthodox economics probably is. One can obviously question the Eurocentric concept of social progress, but there seems to be nothing wrong with a notion of scientific progress as problem solving. Moreover, the unqualified, dismissive attitude toward the ontological or methodological presuppositions of the sciences is also problematic. For instance, Islamists' total rejection of reductionism ignores the spectacular success of physics and chemistry since the seventeenth century. Similarly, reductionism so far also has been successful in many areas of biology and medicine by any standard of success. Unless one can show a better alternative, such dismissive criticism is bound to remain hollow.

Second, we can point out that there is a viable alternative to positivist, social constructivist and postmodernist conceptions of science. Between these extremes is a healthy approach which recognizes that science is one institution among others in a society, and that scientific activity is always carried out in a social context. Scientists are social agents like everybody else, and each has her own prejudices, interests, goals and values in life. It is naive to think that she can simply forget about them when engaged in scientific activity. Various personal and social factors may thus influence scientific work from problem choice to hypothesis formulation. Yet, they can be eliminated through a myriad of ways (such as testing, statistical analysis, analogy, interpretation, peer and self-criticism) if they impede scientific work. In other words, objective scientific activity is possible despite the fact that scientists are social agents.

This suggests that we can accept that science is a socially mediated and value-laden activity without embracing the nowadays fashionable view that science is rock bottom ideology or a game of interest and power. Science is value-laden in two different, epistemic and non-epistemic, senses. By epistemic values I mean predictive and explanatory power, consistency, simplicity, fruitfulness, wide scope and the like. These are values that a good scientific theory must have or aim at. All scientific disciplines are value-laden in this sense, though some values may be more important for some discipline than others. Following Kuhn (1977), I call these factors values, not rules, because their use in theory appraisal requires deliberation and judgment: two scientists who agree on the same set of epistemic values can disagree about their relative significance, or they can disagree in their application to a particular situation. The epistemic values are the product of a long historical process and are in the sole possession of scientific communities. And so far, they have functioned quite well in evaluating scientific theories and hypotheses.

Science is also value-laden in a second sense, namely "what people judge to be good", which involves social, political and ethical values. When Islamists talk about the relationship between science and values, they have this broader sense in mind. Several points need to be made concerning this meaning of values. First, such values are not possessed only by the scientific community; anybody can approach and judge science with them. Second, not all scientific disciplines are value-laden to the same degree in this sense. For example, psychology (with its terms like "normal", "abnormal" and "mental illness") is probably more value-laden than biology, and biology may be more value-laden than physics (compare "adaptation" to "electron"). Finally, and this is the crucial point that we must bring to the attention of the Islamists, the epistemic values can neither be reduced to nor replaced by the non-epistemic ones. No activity can be called scientific if it is deprived of its epistemic values, and no scientific activity would deserve our respect alongside other activities like art, literature and philosophy unless it is sensitive to social, political and ethical values. The middle position I advocate, therefore, emphasizes the social, political and ethical responsibility of the scientist without giving up the relative epistemic-cognitive autonomy of science. As a responsible person, the scientist must continually try to strike a balance between the two kind of values, that is, between producing theories that are simple, consistent, fruitful etc. and at the same time avoiding damage to people, society, and the natural environment. But above all, she should not pretend that science alone can solve all of the problems human beings face in this world, especially the moral ones. We should clearly articulate what science can and cannot achieve, carefully draw its limits and thus cut it down to size (Chalmers, 1990; Graham, 1981).

Third, we can use the classroom, public lecture halls, newspaper columns and TV as convenient forums where opposing theories and worldviews enter into a dialogue with the hope of mutual understanding and change. In spite of the polarization I have mentioned above, such a dialogue is possible and does exist between certain Islamist and other intellectuals, however meager it may be. It is important to keep the dialogue going since the alternative is, as our recent history has taught us, violence between the two camps followed by military intervention. The last time the fight was between "left" and "right"; we must make sure that the next one will not be between "laicists" and "Moslems".

What would be the point of this dialogue? For one thing, Islamist intellectuals may be invited to reflect upon the fact that they are trying to combat the Western form of life, science and thinking (at least in part) with the very same Western rationality and argumentation. Kuhn's historicism, Feyerabend's pluralism, the Frankfurt School and various forms of postmodern thinking are "Western" products. The point is not merely that Islamist intellectuals are being inconsistent, but rather that "West" does not signify an undivided voice, a unified, monolithic way of life and thinking any more than "East" or "Islam" does. There are many different interpretations, attitudes and practices within "West" and "East", "Christianity" and "Islam". Is it, for instance, the Christianity of the Catholics or the Protestants? Is it the Islam of Ziya Gökalp or the radical Islamists, of the Shiites or Sunnies? Instead of dismissing the other as confused, wrong, or evil, one may listen and try to understand. It is only then that we realize that we ourselves are an "other."

One important insight that should emerge from this dialogue is that the criticism of Western civilization that appeals so much to the Islamists is a discourse of *self*-criticism. To appropriate it as simply an anti-Western discourse is to miss the point. The postmodern critique of the authority of science rests on the premise that all claims to universality and absolute truth must be deconstructed. Therefore, neither the "West" nor the "science" which are the objects of this critique can unproblematically be replaced by a supposedly less vulnerable locus of authority such as the "East" or "Islam". Similarly, Feyerabend's attack on "Western" science does not imply the existence of a discourse or form of knowledge that is intrinsically superior to it. The proper response to critiques such as Feyerabend's would have to be a questioning of one's own position of authority.

On the other hand, those who hold a crude positivist view of science and take Western modernity as a paradigm of universal rationality and progress can learn from the divisions within that modernity that there are other and perhaps better conceptions of science, social progress, and rationality – a conception of science which is not positivistic, an idea of social progress which is not Eurocentric, and, finally, a conception of rationality which is not merely instrumental but also substantial. They can learn that the way we acquire knowledge of the world is not independent from the institutions, practices and discourses that produce it and does change historically. The social studies of science can warn against considering the views of the Islamist intellectuals as "unscientific" simply because they are "political", "interested", or "biased."

It seems to me that the most promising attitude is one of tolerance and dialogue. It is based on a spirit of "philosophy in the conversation of mankind" as Richard Rorty put it (1980, p. 389). It may well be our only hope for a peaceful life in a world of increasing instability and deepening conflicts.

References

Armağan, M. (1990). *Islam Bilimi Tartışmaları (Discussions of Islamic Science)*. Insan Yayınları: Istanbul.

Bulaç, A. (1995a). *Din ve Modernizm (Religion and Modernism)*, Iz Yayınları: Istanbul.

Bulaç, A. (1995b). *Çağdaş Kavramlar ve Düzenler (Modern Concepts and Systems)*. Iz Yayınları: Istanbul.

Chalmers, A. (1990). *Science and its fabrication*. University of Minnesota Press: Minneapolis, MN.

Davison, A. (1995). Secularization and modernization in Turkey: The Ideas of Ziya Gökalp, *Economy and Society, 24*: 189-224.

Feyerabend, P. (1987). *Science in a free society*. Verso Publishers: London.

Feyerabend, P. (1988). *Against method*, Revised Edition. Verso Publishers: London.

Feyerabend, P. (1989). Preface to the Turkish Translation of *Against Method*, translated by Ahmet Inam, *Yönteme Hayir*. Ara Yayıncılık: Istanbul.

Feyerabend, P. (1991). Preface to the Turkish Translation of *Science in a Free Society*, translated by Ahmet Kardam, *Özgür Bir Toplumda Bilim*. Ayrıntı Yayınevi: Istanbul.

Foucault, M. (1983). The subject and power. In H. Dreyfus & P. Rabinow (eds.), *Michel Foucault: Beyond Structuralism and Hermeneutics*, Second Edition. University of Chicago Press: Chicago.

Göle, N. (1993). Engineers: 'Technocratic Democracy.' In M. Heper, A. Öncü & H. Kramer (eds.), *Turkey and the West*. I. B. Tauris: London.

Graham, L. R. (1981). *Between science and values*. Columbia University Press: New York.

Gülalp, H. (1995a). The Crisis of Westernization in Turkey: Islamism versus Nationalism. *Innovation, 8*:175-182.

Gülalp, H. (1995b). Islamism and postmodernism. *Contention, 4*:59-73.

Jenkins, E. W. (1992) School science education: Towards a reconstruction. *Journal of Curriculum Studies, 24*: 247-260.

Keyder, Ç. (1987). *State and class in Turkey*. New Left Books: London.

Kuhn, T. (1977). *The essential tension*. The University of Chicago Press: Chicago (320-339).

Kutluer, I. (1985). *Modern Bilimin Arkaplanı (The Background of Modern Science)*. Insan Yayınları: Istanbul.

Lyotard, J. (1984). *The postmodern condition: A Report on Knowledge*. University of Minnesota Press: Minneapolis, MN.

Mardin, □. (1962). *The Genesis of Young Ottoman Thought: A Study in the Modernization of Turkish Political Ideas*. Princeton University Press: Princeton.

Mardin, □. (1981). Religion and secularism in Turkey., In E. Özbudun & A. Kazancıgil (eds.), *Atatürk: Founder of a Modern State*. Hurst: London.

Özdenören, R. (1995). *Müslümanca Düşünme Üzerine Denemeler (Essays on the Moslem Way of Thinking)*. Iz Yayıncılık: Istanbul.

Özel, I. (1988). *Waldo Sen Neden Burada Değilsin? (Waldo Why are You not Here?)*. Çıdam Yayınları: Istanbul.

Özel, I. (1992). *Üç Mesele: Teknik-Medeniyet-Yabancılaşma (Three Issues: Techne-Civilization-Alienation)*. Çıdam Yayınları: Istanbul.

Parla, J. (1990). *Babalar ve Oğullar (Fathers and Sons)*. Iletişim Yayınları: Istanbul.

Parla, T. (1985). *The social and political thought of Ziya Gökalp, 1876-1924*. E. J. Brill: Leiden.

Rorty, R. (1980). *Philosophy and the mirror of nature*. Princeton University Press: Princeton.

Sardar, Z. (1984). *The touch of midas: Science, values and the environment in Islam*. Manchester University Press: Manchester.

Toprak, B. (1993). Islamist Intellectuals: Revolt Against Industry and Technology. In M. Heper, A. Öncü & H. Kramer (eds.), *Turkey and The West*. I. B. Tauris & Company Ltd.: London.

Toprak, B. (1984). Politicization of Islam in a Secular State: the National Salvation Party in Turkey. In S. A. Arjomand (ed.), *From nationalism to revolutionary Islam*. SUNY Press: Albany, NY.

Toprak, B. (1988). The state, politics, and religion in Turkey. In M. Heper & A. Evin (eds.), *State: Democracy and the Military*. Walter de Gruyter: Berlin.

Michael W. Poole

Chapter 9

Science and Science Education: a Judeo-Christian Perspective

As long as we recognize that the world of science is not the *whole* reality we live in, but only one particular aspect of it, and that our religious or our aesthetic experience is as *real* and as *fundamental* as our scientific experience, no harm is done. In that case we will not mistake the scientific experience for the norm of all experience, nor scientific method for the method of all disciplines.
Roijer Hooykaas (1966)

The scientific enterprise, in common with other studies, cannot be pursued without making certain presuppositions which themselves cannot be derived from science. I shall argue that biblically based Judeo-Christian[103] beliefs are congruent with these presuppositions. Historically, such beliefs have provided fertile soil within which Western science has developed and flourished from the seventeenth century onwards (Brooke, 1991; Hooykaas, 1972; Russell, 1985). This favourable environment can be seen as springing from characteristics of the world that might be expected from Judeo-Christian beliefs about the nature of God and the nature of humankind as created 'in God's image'. However, now that modern science has become a mature cluster of disciplines, a diametrically opposite belief has become associated with it in popular thought. It is a belief that turns its back on the origins of Western science within theism and presents science as atheistic. This *volte-face* has been accompanied by various unsuccessful attempts to derive ultimate answers to the meaning of life from science itself. But the failure of science to answer questions about the purpose of life, or how we *ought* to behave, has brought to science its detractors. However, rather than bemoaning science's failure to deliver something which was never in its gift, it could be appropriate to look again for answers to these kinds of questions, to the Judeo-Christian worldview within which science developed and found a prominent place.

A Judeo-Christian Perspective?

One of the first questions that is prompted by the title of this chapter is 'can there be a distinctive Judeo-Christian perspective on science and science education?' One could also ask, 'can there be a secular humanist or an Islamic perspective?'[104], but

[103] Much Christian thinking is rooted in the Old Testament and therefore the issues are similar.
[104] The question of an *Islamic* perspective is raised in Chapter 8 and a *Japanese* perspective in Chapter 7.

W. W. Cobern (ed.), Socio-Cultural Perspectives on Science Education, 181–201.

that is not my subject here. An answer to the first question might start by pointing out that anyone studying disciplines like science or science education will be coming to them with a set of basic beliefs about the world. Such basic beliefs might be that the world is a cosmic accident, or that it is planned; that it is intelligible or that it is not for us to understand. These basic beliefs may bear explicitly or implicitly on the kinds of questions being asked, as well as the kinds of answers that will be counted acceptable. Some of these basic beliefs are deemed open to reconsideration while others are not, for not every belief can be held open to question at the same time. There has to be some stable platform from which other stances can be surveyed and criticised. Even the out-and-out sceptic does not hold scepticism itself open to scepticism at the same time as he is skeptical of all other beliefs. Examples of those beliefs which are not usually deemed open to question are (i) the Cartesian presupposition that, because I can think about these matters at all, I must exist and (ii) the assumption of human rationality, without which meaningful discussions cannot begin. Examples of those beliefs which *are* open to reconsideration are that the universe is unplanned and purposeless, or its converse, that the world is planned and purposeful. People may, and do change these latter beliefs as a result of discussions with theists and atheists. They may also change them as a result of examining the kind of world we live in – an exercise that belongs to the discipline of natural theology.

Worldviews

A person's collection of fundamental beliefs about the world is referred to as their *worldview* (Cobern, 1991). Worldviews in turn give rise to sets of subsidiary beliefs. For example, a person whose worldview includes the belief that the universe is the product of an 'intelligent agent' might have a consequential, subsidiary belief that the world will be intelligible to *us*. Someone who does *not* believe that the universe is the product of an 'intelligent agent' may nevertheless hold, for different reasons, the same subsidiary belief that the world will be intelligible to us, without having an *a priori* expectation that this is likely to be so. Indeed, a belief in the intelligibility of the world can be pragmatically held without reflecting on metaphysical reasons why this might be so. After all, studies of the world, such as are carried out in science, seem to cohere if one *does* make that assumption. It is these shared subsidiary beliefs which make science, as a common cause, possible for those who profess a wide variety of worldviews.

Beliefs and values

There is another connection to be traced before proceeding further, in addition to the one between science and the beliefs that underpin that enterprise. It is the connection between the beliefs which people hold and the things (objects, ideas, traditions) which they value. Values depend on beliefs. The *values* which individuals and societies have are consequences of the *beliefs* that they hold. Those who *value* truth-telling may do so because they believe that, morally, lying is wrong; spiritually, that lying is sinful (i.e., an offense against God); pragmatically, that lying under-mines a society, or for all three reasons. It is evident that, just as similar subsidiary beliefs may arise out of different *worldviews,* so similar *values* may arise out of different beliefs. It is this that makes possible a considerable body of shared values in a pluralist society, without which the society would collapse. The presence of shared values was particularly noticeable in the UK in the findings of the 1996 'National Forum for Values in Education and the Community.'[105] The Forum consisted of 150 people most of whom were nominated by national organisations with concern for young people or education. The Forum agreed that there could be no consensus on:

- the *source* of the values that we all share (some believe that God is the source of all values, others believe that human nature is the only source of values); or,

- how to apply the values that we all share (there are many different ways in which we might show that we value others, for example).

However, it was agreed that a consensus could be reached on the values themselves and that this would provide schools with a basis for the application of these values.[106]

Even where there *is* consensus on 'the *source* of the values that we all share' there will still be disagreements about 'how to apply the values', but where there is no majority consensus on the *source,* the problems of application are compounded. The outcome of a consultation exercise like the one referred to is a pragmatic compromise for a pluralist society in which a variety of different worldviews are extant. In some instances generalisations may conceal widespread differences of interpretation. For instance, in the document's declaration that 'we value families as sources of love and support for all their members and as the basis of a society in which people care for others',[107] the word 'family' is left undefined in a culture which has the highest divorce rate in Europe. The cost of removing ambiguities and aiming at precision and prescription is the attenuation of consensus.[108] Nevertheless, the Forum referred to has made a genuine and valuable attempt to address growing public concern in the UK that there should be an increased emphasis on values – itself an imprecise term.

Pluralism, relativism and truth

The issue of pluralism, which has surfaced here, is frequently associated with relativism. Unlike subjectivism, which locates truth in the mind of the thinker, relativism locates it in the collective decisions of a society. A classical expression of the doctrine of relativism is Protagoras' statement that 'man is the measure of all things', rather than seeing truth as being the way things are in themselves, independently of whether people believe it or not. Although pluralism and relativism are often associated, they are logically distinct. The *fact* of pluralism, that there *are* many different beliefs co-existing in society, does not entail the relativist's claim that

[105] National Forum for Values in Education and the Community, (1996). *Consultation on Values in Education and the Community,* p. 5, London: School Curriculum and Assessment Authority.
[106] *Ibid.,* sections 2.3 & 2.4, p. 6
[107] *Ibid.,* p. 3
[108] The word 'spiritual' is another example of a 'portmanteau word' whose vagueness allows it to be filled up with a variety of different ideas.

there are no absolute truths, even though this assumption is frequently smuggled into discussions about pluralism. A further logical point is that pluralism does not entail the equality of all beliefs, with nothing to choose between them, for if two beliefs involve claims that are mutually contradictory, they cannot both be true. Neither of them maybe true, but both cannot be.

The central assertion of simple relativism is that 'there are no absolute truths; all is relative' – to particular societies. The way in which this assertion 'saws off the branch' on which it is sitting has often been rehearsed. But in view of the fashionableness of relativism, the point bears repeating and it is this: The central assertion of relativism is made because it is believed to be *absolutely true* and consequently it falls foul of the problems of self-reference and reflexivity. The kind of logical muddle one then ends up in can be expressed as follows. If this central assertion of relativism is itself claimed to be true, then it must be *un*true, since it claims that there *is* one absolute truth, namely the claim that 'there are no absolute truths; all is relative'. Furthermore if this claim were true, it could only be relative to particular societies. Hence it could make no claim to universal assent and therefore could not be absolutely true.

Of course beliefs *are* powerfully affected by the societies in which the believers are located, but ultimately the truth or falsity of the beliefs must be tested against the grounds, the evidence, that the beliefs reflect the way the world is.

Both subjectivism and relativism are fashionable, along with the post-modern dislike of *meta*narratives such as the Judeo-Christian worldview. Certainly such stances stand in opposition to key Judeo-Christian beliefs which make universal statements. Such beliefs include God's love for *all* people, his call to *everyone* to repent[109], Christ's death for the sins of the *whole* world and the gift of eternal life to *everyone* who receives God's forgiveness. Universal promises and requirements like these are clearly incompatible with the denial that truth is some kind of correspondence with the way the world is. Equally they cannot be squared with a redefinition of 'true' as 'meaningful' or more vaguely as 'valid'. The point applies whether reference is made to truths about the physical world, such as those with which science is concerned, or to revealed truth, which is part of Christian belief.

What seems to be less clearly recognised is that any onslaught on truth also threatens the scientific enterprise, which has objective knowledge about the world as its goal. Certainly the theory-ladenness of observations, the interpretive factors involved in data handling and the social factors involved in the development of science all need to be taken into account when examining the scientific endeavour. Such factors may make *complete* objectivity unrealisable in practice and raise the question as to how we would recognise it if we achieved it. But that does not rule out objectivity as a goal. The fact that a teacher cannot achieve complete fairness with every pupil, in every class, every day, is not grounds for rejecting fairness as a goal. But once the metaphysical belief that there is an objective world to be reckoned with is denied, the scientific enterprise becomes pointless. If the extreme view is held that

[109] To be distinguished from the popular perception of regret for some past action. Gk *metanoia* (lit. to perceive afterwards, hence to change one's mind in an ongoing response to the will of God).

the findings of science are no more than the outcomes of social conditioning and do not reflect, however imperfectly, the world as it is, what is the point for its practitioners? If such a view were to be widely believed by the scientific community, it is not clear that the scientific enterprise could be sustained as an activity of society.

Even less clearly perceived in some quarters, it seems, are the consequences for those people who make this extreme claim that scientific knowledge is simply the result of social conditioning. Such a claim has a devastating entailment. This is that any social studies of the social conditioning of scientists will themselves be simply the result of social conditioning, rather than a reflection of what scientists objectively do. This in turn raises questions about the value of the results of such studies, if they are no more than the results of the social conditioning of those students of society. The matter of the truth or falsity of what is believed about God, the natural world and about societies remains a key issue for religious believers, scientists and sociologists alike.

The fact of pluralism raises the question of how what I have said, and shall be saying, about Judeo-Christian perspectives on science relates to other world faiths. The answer is that where belief in a transcendent, rational God is held, as for example in Islam, several of the points that I am making also hold good. Islam has made particularly noteworthy contributions to the development of science from the eighth to the fifteenth centuries and today there is a revival of Muslim interest in relating science to an Islamic worldview.[110] Chinese science, too, has made important contributions from very early times and the Marxist historian, Joseph Needham (1969), has made a magisterial survey of these contributions. It is of particular interest to our current concern with how metaphysical factors affect the development of science that he poses the following question and offers the subsequent answer:

> Why did modern science, the mathematization of hypotheses about Nature, with all its advanced technology, take its meteoric rise only in the West at the time of Galileo? (p. 16)
>
> [But in China] the available ideas of a Supreme Being, though certainly present from the earliest times, became depersonalized so soon, and so severely lacked the idea of creativity, that they prevented the development of the conception of laws ordained from the beginning by a celestial law-giver for non-human nature. Hence the conclusion did not follow that other lesser rational beings could decipher or reformulate the laws of a great rational Super-Being if they used the methods of observation, experiment, hypothesis and mathematical reasoning. (p. 37)

The preceding comments have begun to suggest some answers to the anticipated question about why a Judeo-Christian perspective or perspectives on science might be expected. This serves as a necessary prelude to seeing how such a perspective might affect science education, with its concerns for teaching about the nature of science, its content, how it is used in a social context and how it relates to other subject areas in the curriculum. So the next three steps towards developing an

---

[110] Conference Papers of the International Conference on Science in Islamic Polity in the Twenty-first Century (March 26-30, 1995) Islamabad-44000 Pakistan: OIC Standing Committee on Scientific and Technological Cooperation (COMSTECH).

answer to our question are to trace out those basic beliefs about the world which are needed in order to do science; to sketch out Judeo-Christian beliefs about the world; and, to see how Judeo-Christianity might provide a set of beliefs favourable to the practice of science.

(i) Beliefs underpinning the scientific enterprise

No one would undertake the practice of science unless they believed that,

- human reason is generally reliable,
- there is regularity and order in the universe,
- humans can discover and understand something of that order,
- there is a basic uniformity in the behaviour of the natural order, in space and time. This is expressed in two ways. First, the laws of nature are the same throughout space and have remained unchanged through time (with the likely exception of the first fraction of a second of the early universe). Second, present-day processes provide keys to unlock the secrets of the past. And,
- science is a worthwhile activity.

(ii) Judeo-Christian beliefs about the world

In spelling out some of the fundamental beliefs of the Judeo-Christian position, it is not necessary to give a comprehensive theology, but only to rehearse those aspects that are relevant to the scientific enterprise and to science education. Orthodox Jews and Christians believe that the world is created by God;[111] is maintained in being, moment by moment, by divine activity; has a purpose; was planned and is not accidental.

As every teacher of religious education will have found, reference to God as a 'person' frequently conjures up in young people – and older ones too, anthropomorphic images of larger-than-life human beings. This, together with the literal perception of metaphoric language, prior to the developmental stage of formal operations, has the unfortunate consequence of generating a range of inappropriate and often bizarre mental images of God. J. B. Phillips (1969, p. 5), in *Your God is Too Small,* lists more than twelve of these pictures, including Resident Policeman, Parental Hangover, Grand Old Man, God-in-a-Box and Resident Director. But the employment of metaphors and models is a consequence of the limitations of language, as well as the limitations of our minds, when trying to be articulate about the novel, the unseen and the conceptually difficult.

---

[111] As soon as the word *God* is mentioned the limitations of language are evident. How should one refer to God? The Old and New Testaments clearly indicate that personal language is the most appropriate, so we loosely refer to God as a 'person', or rather, as expressed in Trinitarian theology, as three 'persons' constituting the Godhead. These three are, of course, Father, Son and Holy Spirit. Furthermore, as personal language involves gender, it raises another aspect of the question as to how to refer to God. The Bible uses both female and male metaphors about God, although male ones predominate. God's love and protection is compared to a hen sheltering her chickens and to a shepherd caring for his sheep. But God's protection is also referred to as a 'strong tower', which is non-personal. God is not male or female and questions about God's gender are inappropriate. Since a choice has to be made about how to refer to God, or else to remain silent, I shall adopt the convention of using 'he', without implying any negative thinking about gender.

> ...it is a serious mistake to think that metaphor is an optional thing which poets and orators may put into their work as a decoration and plain speakers can do without. The truth is that if we are going to talk at all about things which are not perceived by the senses, we are forced to use language metaphorically. Books on psychology or economics or politics are as continuously metaphorical as books of poetry or devotion. (Lewis, 1947, p. 88)

Metaphoric language is indispensable in religion as it is in science. The appropriateness of the figures of speech which are used is of central importance and this is an area of intense interest within the philosophy of religion (Soskice, 1985). The idea of '*creation*' has several strands associated with it, including those of:

- ***'bringing-into-being'*** – Something new is now in existence, something that is declared 'good' in the book of Genesis, and later said to have been damaged on account of human sins of commission and omission. 'Creation', in this sense of 'bringing-into-being' is a word frequently 'borrowed' from theology and applied to ideas, fashions and objects. It is also widely used in cosmology, but without theistic connotations, to refer to the coming-into-being of the universe.
- ***ownership*** – with its associated rights. Books commonly contain the following rubric: 'The author asserts the moral right to be identified as the author of this work.' This carries with it a responsibility for how the work is used.
- ***relationships*** – in addition to ownership, since some of that which is created, notably humankind, is not simply owned as an object, but can consciously respond.
- ***distinction between the creation and the Creator*** – The two are not one and the same, as is held by *pantheism.*
- ***an act*** – which is to be distinguished logically from whatever processes may have been involved. Thus, to try to contrast the *act* of creation with the *processes* of evolution is to commit some kind of what Gilbert Ryle called a *category-mistake.*[112]

The idea that God sustains the universe in a moment-by-moment way is often included under 'creation', since 'sustaining' can be viewed as an aspect of the ongoing relationship between the Creator and the creation. Nevertheless, it is helpful to list it as a separate aspect of divine activity in order to distinguish between orthodox Judeo-Christianity and deism. Deism, of which there are many forms, portrays God as the Cosmic Clockmaker, the Retired Architect, the Absentee Landlord who, having brought the world into being, removes from the scene, leaving the creation to its own devices. Orthodox Judeo-Christianity presents God as 'sustaining all things by his powerful word'[113], i.e., it is not simply that if God were

---

[112] A *category-mistake* is made if what can only appropriately be said about something in one category is said about something in another. See Ryle, G. (1963) *The Concept of Mind,* pp. 17f, Harmondsworth: Penguin. In this instance, *acts* and *processes* are in different categories. The oddity of the commonly accepted assertion, 'Either God created the world or it happened by evolution', becomes apparent when a sentence with a similar logical form, also embodying an *act* and a *process,* is substituted. Thus, 'Either the Ford team created the new car or it happened by automation' reads very strangely.

[113] Hebrews 1:3 (Bible quotations, with the exception of Psalm 111:2, cited later, are from the New International Version)

not active the universe would look rather different, but rather that there would be no universe. Thus, whereas deism holds that God is *transcendent* (greater than, other than) it denies his *immanence* (his ongoing involvement in the world). *Pantheism*, by identifying God with the natural world, holds that God is *immanent* but denies his *transcendence*. Traditional Judeo-Christian theism holds that God is both immanent and transcendent.

Table 1 offers a summing up the interrelationships between *pantheism*, *deism* and *theism* and the concepts of *transcendence*, *immanence* and *nature*.

Table 1

	Pantheism	Deism	Christian Theism
Transcendence	—	God as retired architect	God is 'other than'
Immanence	world = God	—	God 'involved with'
Nature - deified	deified	de-deified; deification regarded as idolatry	

The idea of *purpose* is associated with creation and sustaining in Judeo-Christian theology. Purpose is linked with the concepts of intentions, aims, goals and with one use of the word 'meaning'. Biblical teaching about ultimate purposes involves the forgiveness of sins through Christ's death on the cross for all who will receive it, and the redemption of the whole world. Its verdict on the world and its promises for the future are spelt out thus:

> ...the whole creation has been groaning as in the pains of childbirth right up to the present time.' Also, '...the creation itself will be liberated from its bondage to decay and brought into the glorious freedom of the children of God.[114]

So, how does this idea of purposefulness relate to the scientific enterprise? Does it for instance link up in anyway with the different scenarios which science presents for the possible future of the physical universe? I hesitate to think that it does, for the passage is primarily concerned with the purposes of God. However, since the future of the universe itself constitutes part of those purposes, room is left for debate. But even raising the question highlights the differences between the respective concerns of science and theology. Answers to questions about the overall meaning and purpose of 'life, the universe and everything', are not derivable from science whose concerns are with mechanisms rather than meaning. Indeed, some scientists have gone to great lengths to try to purge science from any kind of teleological (goal-directed) language. So, to claim, for example, that the reproduction of DNA is the purpose in life, is to use language in an odd manner.[115] Such a claimed purpose

[114] Romans 8:22 &21

[115] Many people see function as reflecting purpose, or at least use that sort of language — 'the purpose of chlorophyll is to carryout photosynthesis'. To claim that the reproduction of DNA is the purpose in life is to talk about the *function* as if it were a *purpose*.

certainly does not reside in the intentionality of the world, which does not possess a mind. Neither is intentionality involved at the level of the DNA, which possesses no mind that would enable it to plan its own replication. On evolutionary theory, if replication takes place, life is perpetuated, if it does not, life dies out. One could, however, speak coherently of the replication of DNA as being *one of* the purposes of a divine agent (as distinct from DNA), as an integral part of his overall purpose.

Summarising this section, in Judeo-Christian theology the three roles of Creator /Sustainer/Redeemer are associated with the agency of Christ, the second 'person' of the Godhead, in two quite explicit passages shown in Table 2 below.

(iii) Judeo-Christian attitudes to science

While modern science was developing within a Judeo-Christian culture, it was commonly believed that the Bible provided incentives for doing science. For example:

- The biblical portrayal of a rational, orderly and non-capricious God encouraged a degree of confidence in human reason and a belief that there was an order in nature, without which science would be impossible. In the Bible God is portrayed as reasoning — 'The *reason* the Son of God appeared was...' [I John 3:8]; people are encouraged to use their reason — 'Always be prepared to give an answer to everyone who asks you to give the *reason* for the hope that you have.' [I Peter 3:15] and God is said to challenge people to use their reason — ' "Come now, let us *reason* together," says the LORD.' [Isaiah 1:18]

Table 2

He is the image of the invisible God, the firstborn over all creation. For by him all things were created: things in heaven and on earth, visible and invisible, whether thrones or powers or rulers or authorities; all things were created by him and for him. [*Creator*] He is before all things, and in him all things hold together. [*Sustainer*] And he is the head of the body, the church; he is the beginning and the firstborn from among the dead, so that in everything he might have the supremacy. For God was pleased to have all his fullness dwell in him, and through him to reconcile to himself all things, whether things on earth or things in heaven, by making peace through his blood, shed on the cross. [*Redeemer*] (Colossians 1:15-20)	In the past God spoke to our forefathers through the prophets at many times and in various ways, but in these last days he has spoken to us by his Son, whom he appointed heir of all things, and through whom he made the universe.[*Creator*] The Son is the radiance of God's glory and the exact representation of his being, sustaining all things by his powerful word. [*Sustainer*] After he had provided purification for sins, he sat down at the righthand of the Majesty in heaven. [*Redeemer*] (Hebrews 1:1-3)

- Judeo-Christian beliefs also encouraged an expectation that nature would be intelligible, since humans are made in God's image. The idea of being made in God's image, about which volumes have been written, includes those attributes of personality such as moral responsibility, rationality, knowledge of God and God's will for humankind (through revelation), managerial responsibility and creativity. In line with this expectation that the world will be intelligible, the famous 'Research Workers' Text', Psalm 111:2, carved in Latin and English respectively, over the entrances to both the old, and the new Cavendish Laboratories in Cambridge, UK, declares, 'The works of the Lord are great, sought out of all them that have pleasure therein.'
- The basic uniformity in the behaviour of the natural order is expressed in passages like Ecclesiastes 1:5,7:

> The sun rises and the sunsets, and hurries back to where it rises... All streams flow into the sea, yet the sea is never full. To the place the streams come from, there they return again.

- The charge of managing the creation (Genesis 1), 'subduing' and 'ruling over' (but not abusing) the earth, appeared to license science as a way of fulfilling that trusteeship.
- It seemed that God could be glorified through gaining knowledge of the world which could (i) help relieve suffering; (ii) reveal God's power and wisdom.

Both of these, as well as the command to manage the creation, encouraged the view that science was a worthwhile activity. Point (ii) was in turn strongly linked with awe, wonder and the worship of God in passages like Psalms 19:1 & 8:3f:

> The heavens declare the glory of God; the skies proclaim the work of his hands.
>
> When I consider your heavens, the work of your fingers, the moon and the stars, which you have set in place, what is man that you are mindful of him, the son of man that you care for him?

John Brooke (1991, p. 18), historian of science, observes,

> There may be no obvious equivalent in science of the call to worship in religion. And yet, there were those in the late seventeenth century, Robert Boyle (1627-91) and John Ray (1627-1705) among them, who envisaged scientific enquiry itself as a form of worship. The image of nature as temple, the scientist as priest, was explicit in Boyle.

- Creation was viewed as a free act of God, *contingent* (did not have to be that way) rather than *necessary*, so one needed to go and do experiments to find out about the world, rather than appealing to reason alone, or to authorities like Aristotle.
- Those ancient Greeks who identified God with the world (pantheism) saw experimentation on a semi-divine nature as *hubris*, as intellectual arrogance above one's station, something akin to sacrilege. This discouraged experimental science. Judeo-Christianity, with its belief in a God who was distinct from creation, although involved with it, removed that obstacle. The danger of intellectual arrogance was still real, however, but it was not against a putative

semi-divine Nature, but against God. So when, in Chapter 38 of the book of Job, Job becomes rather 'too big for his boots', an uncomfortable conversation ensues, in which God asks:

> [2] 'Who is this that darkens my counsel with words without knowledge? [3] Brace yourself like a man; I will question you, and you shall answer me. [4] Where were you when I laid the earth's foundation? Tell me, if you understand... [12] Have you ever given orders to the morning, or shown the dawn its place...? [18] Have you comprehended the vast expanses of the earth? Tell me, if you know all this... [21] Surely you know, for you were already born! You have lived so many years!... [33] Do you know the laws of the heavens?'
>
> [Two chapters later] [3] Job answered the LORD: [4] 'I am unworthy – how can I reply to you? I put my hand over my mouth. [5] I spoke once, but I have no answer – twice, but I will say no more.'

Additionally, cases could be made out that the criteria of elegance and parsimony of causes for the selection of scientific theories might also be reflections of God's nature.

From the foregoing it emerges that two domains of issues are involved, *affective* and *cognitive*. The affective issues include awe, worship and humility. The cognitive issues include the presuppositions of rationality, orderliness intelligibility and uniformity. Some issues, like moral responsibility and helping to relieve suffering, combine both domains. Indeed, an example of the interplay between the two domains is furnished by the point already made, namely that many scientists, past and present, have said that their scientific work was motivated by their commitment to their Creator God. They saw their science studies as a form of worship as they sought to understand God and his world better.

The ghost of the 'God-of-the-gaps'

What I have sketched out above is a synthesis of what I regard as a coherent view of overall biblical teaching which is relevant to the scientific endeavour. For the sake of completeness it needs to be said that not all Christians have seen science in a favourable light. Some have, and still do, see it as a threat. Adopting a 'God-of-the-gaps' mentality they have seen each new scientific explanation as threatening to oust God from the scene. By failing to distinguish between the type of explanation involved in the scientific chain-mesh of proximate causes and the type of explanation concerning the agency of God, they have committed an explanatory type-error and caused themselves unnecessary distress. A moment's reflection reveals the oddity of a position which imagines scientific explanations as displacing the actions of agents. Would those who hold such a view concerning God seriously entertain the idea that a scientific explanation of a hovercraft would mean one could no longer believe that the creator of this unique form of transport was Frank Cockerell?

This failure to distinguish between different types of explanation is not confined to the religious believer. There are those who, while not believing in God themselves, believe that the idea of God entails a 'God-of-the-gaps' mentality.

Accordingly, they may pursue science with additional enthusiasm, thinking thereby to close any gaps where God might be imagined to lurk. The distress of the first group and the enthusiasm of the second are based on the same insecure foundation of a monolithic view of explanation.

There are still other attitudes to science taken by Christian believers. Some take a non-interactionist view by placing science and religion in separate watertight compartments. I regard such a position as being difficult to defend in view of the many areas of study in which interactions have taken place. In brief, these interactions can be grouped in three main ways that involve the *data*, the *nature* and the *applications* of science. Table 3 below lists many of the main issues that fall into these three categories. The sheer range of issues has generated whole libraries of books that examine them. I have discussed a number of them at length elsewhere (Poole, 1995), but here I simply wish to make a brief comment under each of the three headings.

Table 3

Data	Nature	Applications
Origin of: • the universe (stellar evolution) • life (chemical evolution) • species (organic evolution) the Earth • position (Galileo affair) • status • age human nature mind and brain	scope, limitations & status of science — logical positivism determinism, freewill & chaos theory miracles, scientific laws and the uniformity of nature orderliness, causality explanations, reductionism & emergence language — analogies & models creation evidence & proof facts and faith, faith and reason	medical ethics ecology energy use and power generation military health and safety in industry and at home

The data of science: Many interactions have taken place between the discoveries of science and statements in the Bible. Sometimes these have seemed to present problems, as with the nineteenth century age-of-the-earth and evolution debates. But a large number of these problems have undoubtedly arisen through trying to read modern scientific ideas out of ancient Hebrew texts and failing to recognise the appropriate literary *genre* involved. Sometimes the interactions have offered interesting insights into the biblical narrative, as for instance over the 'star' of Bethlehem (Humphreys, 1991). On other occasions they have raised questions about possible purpose and design over the apparent 'fine-tuning' of the universe for life to be present.[116]

The nature of science: The nature of the scientific enterprise has sometimes been thought to cast doubt on matters like miracles (in view of the assumption of the uniformity of nature) and also to raise questions about human freewill (on account of the idea of determinism).

[116] The so-called Anthropic Cosmological Principle.

The applications of science: Some people, seeing the negative uses to which science and technology can be put, have forgotten its positive benefits and regarded it as anti-God – 'of the devil'. Others have adopted the blinkered approach of regarding science as something that disturbs the world-as-God-intended-it – if God had intended us to fly he would have given us wings. Food for thought is provided by the words of the old lady referred to by the comedians, Flanders & Swann, 'if God had intended us to fly he would never have given us the railways'! As will be realized from what I have already said, I believe these abreactions to science, supposedly in the name of Judeo-Christianity, to be profoundly wrong. Yes, of course science has been misused; so have steel, sex and speech. But the antidote to abuse is right use, not disuse. The solution to the problem of drunken driving is not to abolish the car.

If the misuse of science can bring bondage, its careful use can bring new freedoms. Many years ago, the Dutch historian of science, Reijer Hooykaas (1957, pp. 8f) wrote of the need for 'both an outward liberty of science and also an inner liberty', arguing 'that the inner freedom necessary to scientific work is fully *guaranteed* by a biblical religion.' He acknowledged 'how much a certain exegesis of biblical texts has hindered the free development of scientific theory' but pointed out that 'The first effect of the inner liberation brought about by the biblical message, however, should be that we fully recognise that good Christians may be bad scientists, poor exegetes and silly people.'

The data, nature and applications of science are all important to science education. In the *data of science* domain, a careful treatment of the historical interactions between science and Judeo-Christian belief over such matters as the Galileo affair and the Darwinian controversies can help to dispel some of the dubious folklore surrounding these episodes. Such folklore has attracted the attention of historians of science who have investigated how these folklore accounts have been used to promote and perpetuate the misleading 'conflict thesis' about science and religion. In the area of science education there has been an increasing interest in teaching about the spiritual, moral, cultural and social contexts of science. In the *nature of science* domain, a study of the limitations as well as the strengths of science is helpful in dispelling the logical positivists' notion of science as the ultimate source of truth, a notion that is still deeply entrenched in popular thought. It is important that pupils should understand the philosophical deficiencies of such an imperialist view of science. This episode in the history of recent philosophy is also important in helping pupils to understand how science came to be promoted by various pressure groups as anti-religious, something that was claimed to follow from a now-discredited view of science. Most of the points about the nature of science which have featured in interactions with religious belief can be argued as needing to be treated in the teaching of science, simply as part of good science teaching, irrespective of their connections with religious beliefs. In the *applications of science* section, the role of science is confined to its competence to indicate the likely consequences of particular courses of action, not whether the actions themselves are morally right of wrong. The applications of science is an area of particular importance to Judeo-Christianity, with its concern for the rectitude or otherwise of actions and their relation to divine commandments.

In this section, a number of the things which I have said, including some of the beliefs referred to, are not exclusive to Judeo-Christianity. My main point is that the beliefs necessary in order to engage in the scientific enterprise are thoroughly congruent with Judeo-Christian teaching and also that, where Judeo-Christian beliefs are held, there exists fertile soil for the nurture and development of science.

Ethics, metaphysics and science

In the earlier consideration of Judeo-Christian incentives for engaging in science it emerged that two domains of issues are involved, the affective and the cognitive. The affective ones included ethical issues such as how we ought to behave. The cognitive ones included metaphysical prere-quisites for science, like rationality and intelligibility. Although such issues are not derivable from science itself, they have been the subject of vigorous attempts to find in science itself the source of the moral imperatives for living and, at the cognitive level, the source of its own justification. With respect to attempts to derive ethical systems from science, the history of science records numerous attempts to bridge the 'logical Grand Canyon' (Flew, 1970, p. 39) between what IS the case and what OUGHT to be done. A major example is the attempted derivation of Evolutionary Ethics, with which the name of the late Sir Julian Huxley is particularly associated.

The history of science records another example of the interplay between science and ethical ideas, one that is completely opposite in its intention to the attempts, referred to above, to establish evolutionary ethics. Here the aim was not to derive ethical systems from science, but to use science to *excise* ethics (along with religion and metaphysics in general) from the realm of meaningful discourse. The scene was set in Vienna in the 1920s and 30s and concerned a group of philosophers known as the Vienna Circle. Their philosophy, known as logical positivism, referred to above, elevated the 'language of science' to the status of a *meta* language, against which all other claims to meaning had to be judged. They allowed that ethical talk was an expression of emotion, such as 'cheating — ugh!', but claimed that it was cognitively meaningless. The history of the rapid rise and subsequent demise of this philosophical position are widely known and I have spelt out its implications for science and science education elsewhere (Poole, 1995, p. 33ff). Suffice it to say that the position contained the seeds of its own down fall; for if it were true, then logical positivism itself fell an early victim of its own philosophy. By insisting on empirical proofs for meaningful statements, other than those like definitions that were analytic, it rendered its own Verification Principle meaningless. What was particularly interesting as far as the present subject is concerned is that this episode marked a peak in the *hubris* which one view of science had taken on. With its disdain for metaphysics, logical positivism was nevertheless shown to be a metaphysical position.

Science, both (i) through one attempt to make it the source of moral 'oughts' and (ii) through another, opposite attempt to remove moral 'oughts' from the realm of meaningful discourse, had mutated suddenly to produce a 'sport' – 'scientism' – which turned its back on its rapid growth within Christendom and tried to bite the hand that reared it. Coupled with a nineteenth century disenchantment with the

Established Church and the struggles for professionalism in science, science and 'scientism' were often not differentiated. 'Scientism', however, also produced its mutants, and some of these were mutually contradictory, as indicated above. These variants of scientism, however, are not so much distinguished by what they affirm, but by what they deny, and Table 4 below paints with a broad brush some positions that are typical of scientism.

Table 4

Science	Scientism
is the study of nature (Gk. *phusis* - physics); *methodologically* excludes metaphysics[117]	Denies that anything other than the natural world exists – *naturalism* – rejects metaphysics
(i) reduces systems to their components – *analytic* – methodological reductionism (ii) also recognises the *emergence* of higher order properties with increasing complexity of organisation – *synthetic*	Scientific accounts are all there are – ontological[118] (metaphysical) reductionism
Causality – as a methodological agreement science only deals with *efficient* causes	Denies that there are *first* causes or *final*[119] causes (teleology)
(i) utilises the fruitfulness of a machine metaphor, while recognising its limitations (ii) may employ an organismic metaphor e.g. for the environment	(i) world is nothing more than a machine (ii) may vest nature with godlike properties – *Gaia* and embrace *pantheism*
Scientific laws describe the normal behaviour of the natural world. The uniformity of nature is a methodological assumption	Denies that there could ever be behaviour other than law like (anti-miraculous)
one form of knowledge, differing from historical, personal, mathematical etc. forms	the only – or at least the best – form of knowledge
assumes rationality in the interpretation of empirical evidence	Exalts reason and the empirical and denies revelation
Deals with the processes of *evolution, natural selection* and *chance* but avoids metaphysical extrapolations	Reifies[120] concepts such as *nature, chance, evolution and natural selection* and vests them with God-like attributes

Scope and limitations of science

There continue to be those who see science as omnicompetent to answer every kind of question. But will science ever be in a position to give a complete account of the

[117] The term *metaphysics* was originally used as a title for those books that came after Aristotle's *Physics*. Now it is used to refer to enquiries that raise questions about realities that lie beyond the capabilities of science to answer.

[118] Ontology is about the nature of being, with what exists.

[119] *Final* causes, in Aristotle's fourfold notation, are the goals towards which the results are directed.

[120] Treats a concept as though it were a real object or cause.

universe? A moment's reflection suggests that there are in fact two distinct questions involved, rather than one. First, will it ever be in a position to give a complete account, in physical cause-and-effect terms, of the material world that is its province? Second, will it ever be in a position to give a complete account of the universe in its totality, including morals, metaphysics and religion? We will consider the two questions under the headings of 'Theories of Everything' and 'The omni-competence of science?'

Theories of Everything: The 'Holy Grail' of some scientists is to find a 'Theory of Everything [TOE]', 'a single all-embracing picture of all the laws of Nature from which the inevitability of all things seen must follow with unimpeachable logic' (Barrow, 1991, p. 1). It must be made clear from the start that careful writers like Barrow point out that 'we must be circumspect in our use of such a loaded term as 'Everything'. Does it really mean everything: the works of Shakespeare, the Taj Mahal, the Mona Lisa? No, it doesn't' (Barrow, 1991, p. 2). But, even with this proviso, the enterprise seems fraught with difficulties for a variety of reasons. Various writers, including Barrow (1991) and Trigg (1993), have indicated some of the difficulties of the enterprise.

- It seems essentially reductionistic, physics-based and does not obviously account for every aspect of the phenomenon of emergence — the appearance of novel properties at higher orders of complexity on account of the arrangement of the constituent parts.

- The search for a mathematical formulation of a theory appears to beg the question as to why mathematics might be expected to mirror reality. As Stephen Hawking (1988, p. 174) puts it:

  > What is it that breathes fire into the equations and makes a universe for them to describe? The usual approach of science of constructing a mathematical model cannot answer the questions of why there should be a universe for the model to describe. Why does the universe go to all the bother of existing?

- If a Theory of Everything *were* complete, how would we know that it was complete?

- Gödel's theorem suggests that any explanation which claimed to be a Theory of Everything would be unable to explain its own existence.

- Heisenberg's Principle of Indeterminacy tells of a fundamental limit to what we can know about a physical system. Although the Principle applies to all systems, its practical effects are most noticeable at the microphysical level. Any attempt to find the position of something like an electron requires a minimum interaction with a quantum of energy. This interaction acts as a kind of 'kick' which changes the electron's *position* and its *momentum*. Consequently the electron is no longer *where* it would have been, nor moving with the *momentum* it would have had, if we had not been observing it. This constitutes a theoretical limit to the completeness of scientific predictions. Now, it could be argued that a Theory of Everything *might* predict such an indeter-minacy principle. But referring back to Barrow's definition of a Theory of Everything as 'a single all-embracing picture of all the laws of Nature from which the inevitability of all things seen must follow with unimpeachable logic', the predicted *indeterminacies,* as distinct from the predicted *Principle*, remain as a limit on further predictions, so the future does *not* proceed with 'unimpeachable logic'. Although, in order to do science, one assumes a basically deterministic system as a methodological principle, nevertheless at the microphysical level, deter-minism seems to be violated.

- Chaos theory has indicated that some physical systems, even though taken to be deterministic systems, are exquisitely sensitive to minute differences in the initial conditions. Consequently, the smallest differences at the outset can in time produce very large changes in the outcomes. This phenomenon is popularly known as the 'butterfly effect' due to the suggestion that the flapping of a butterfly's wings on one side of the world may in time trigger a storm on the other. Once again it could be argued that a claimed Theory of Everything might indicate that this sensitivity would exist, but the existence of this sensitivity still places a limit on possible scientific prediction. Once again, the future does not proceed with 'unimpeachable logic'.
- Other limitations on our knowledge of the world and on our ability to predict are posed by certain intrinsic properties of the universe itself, when extensive computation is involved. According to Stockmeyer and Chandra, certain computational problems which are solvable in principle run up against limits in the ultimate size and speeds of computers:

> The most powerful computer that could conceivably be built could not be larger than the known universe (less than 100 billion light-years in diameter), could not consist of hardware smaller than the proton (10^{-13} centimeter in diameter) and could not transmit information faster than the speed of light(3×10^{8} meters per second). Given these limitations, such a computer could consist of at most 10^{126} pieces of hardware ... this computer, regardless of the ingenuity of its design and the sophistication of its program, would take at least 20 billion years to solve certain mathematical problems that are known to be solvable in principle. Since the universe is probably not 20 billion years old, it seems safe to say that such problems defy computer analysis. (Stockmeyer & Chandra, 1979, p. 124)

Many of the points raised above are still subjects for discussion and debate. But they should serve as cautions against grandiose claims about the scientific enterprise. Their main importance here, from Judeo-Christian (and many other) standpoints, is to emphasise that humility is appropriate to our finitude when confronted by this remarkable universe. The points to be raised below, however, are of a different kind. Instead of being about the possibility of completeness of science, they question the competence of science to answer all the legitimate and meaningful questions which people ask.

The omnicompetence of science? In considering the question of whether science will ever be in a position to give a complete account of the universe in its totality, including morals, metaphysics and religion, it is important not to confuse the question with the kind of 'God-of-the-gaps' thinking already referred to. A 'God-of-the-gaps' approach involves the explanatory type error of mixing together two distinct types of explanation — reason-giving explanations of agency and motives on the one hand and reason-giving explanations of physical cause-and-effect on the other. This results in trying to slot 'God' into gaps in scientific knowledge, gaps which ought to be filled with scientific explanations as they become available and *not* with talk-about-God. Here, however, the issue is one of a straightforward recognition that there are different, but compatible, types of explanation which answer different explanation-demanding questions. To express the limitations of science in another way, it is not that there are limits on the territory which science is able to explore, although there may well be moral restraints on some kinds of research. Rather it is that the limitations of science lie in the kinds of questions that can be answered by the methods it employs. As has been pointed out, the limitations of science are methodological, not territorial.

As with Theories of Everything, claims for the omnicompetence of science in all domains of enquiry seem fraught with difficulties of their own:

- If science is the study of physical world, of nature, there appears to be a logical impasse in expecting it to be able to pronounce on religious questions of the kind 'is there anything *other* than nature (God) to which nature owes its existence.
- Attempts to make science, in the form of evolutionary biology, a source of moral imperatives run up against the Is/Ought divide. In an old, but remarkable book, *Recovery of Belief,* the former atheist philosopher, C. E. M. Joad (1952), tackled the issue of evolutionary ethics as part of his 'account of some of the reasons which have converted me to the religious view of the universe in its Christian version' (p. 13). He asked:

> what sort of ethical criterion could the process of evolution itself provide?...It tells us, in fact, what has occurred but it does not tell us, nor could it do so, what *ought* to have occurred. In informing us that certain events have taken place, it cannot and does not assure us that it is *desirable* that they should have taken place... the types which survive are deemed to be valuable by no criterion other than that of the fact of their survival. (Joad, 1952, p. 108)

Evolutionary theory, then, can only offer an explanation of the development of some aspects of behaviour which might be counted moral in its own limited terms of survival value. Altruism is one such example. But since Hitler's goal of breeding a master-race was based on evolutionary ideas of survival, it can be seen that morally right behaviour cannot be judged simply on the consequences of survival. It can always be asked whether survival itself is a good thing. It has even been suggested that human beings are a 'cancer on the globe'.

Evolution can also offer a restricted account of the development of those aspects of rationality that aid survival, an area of study known as evolutionary epistemology. But it is far from clear that evolutionary epistemology can be extended to account for the whole complex range of human cognitive processes and creativity. Evolutionary theory does not provide a secure base from which to argue for the primacy of science, by claiming that science can provide an ethical system as well as physical knowledge of the world. Conversely, it is no more proper to exalt science to omnicompetence by claiming, as did the logical positivists, that science renders ethical talk cognitively meaningless. The primacy of science is not secured by a stipulative definition that asserts that all other claimants for a place are meaningless entities.

Finally, science does not answer the question, 'why is there something rather than nothing?'

Judeo-Christianity reconsidered?

I believe the Judeo-Christian view of science, which I have tried to sketch out, is a high one. It stands against both the despising of science and its deification. The great power of science requires great responsibility – to other human beings and also to God. But although science stands high in the list of human activities, it has its limitations and needs to be pursued with humility. We know very little of the totality there is to know and it seems that the more we unpeel the 'cosmic onion' the more layers there are to be found. Viewed in Judeo-Christian perspective, science is a major human activity, but it leaves out by its very nature those questions that in every age are asked, questions like, What is the purpose to life? Do I have any importance in this vast universe? What happens when I die? And, is life 'a tale told by an idiot, full of sound and fury, signifying nothing'?[121]

I referred earlier to Joad's book, *Recovery of Belief*, and even as I write, my eye catches another title on the bookshelf, *Science and the Renewal of Belief* by Stannard (1982). The word 'renewal' is indicative of a growing interest in the interplay between science and religious belief. Following the recognition within *academia*, if not among the general public, of the inadequacy of the so-called *conflict thesis* as a description of the complex interplay between science and religious belief, there has come a new liberation in studies about the interactions. The *Who's Who in Theology and Science*[122] testifies to a world-wide, burgeoning academic interest in this area.

I have made clear my own commitment to the Christian faith, although much of what I have said also applies to other major theistic religions. Within this Christian worldview I regard science as a fascinating human activity, compatible with biblical teaching. This does not mean that there are no problems or issues still to be debated and thought through. Such issues will always exist. What I have written here reflects my current thinking, but that will continue to be modified as new data and new ideas come along. No doubt there will be many more times when I shall have to admit errors. That is an element of being involved in any academic enterprise and is part of

---

[121] William Shakespeare's *Macbeth*, Act V, Scene 5

[122] *Who's Who in Theology and Science, (1996)*, John Templeton Foundation

what makes it dynamic and interesting. But, in my view, it is still possible to hold to a properly circumscribed version of the 'God's Two Books' metaphor (Poole, 1995, p. 77f) seeing a common author to the Book of Scripture and the Book of Nature — the Book of God's Words and the Book of God's Works. From the second of these two 'books' come the insights and wonders of a world of breathtaking complexity, of which the astronomer, Sir James Jeans (1930, p. 148) once said,

> ...the universe begins to look more like a great thought than a great machine. Mind no longer appears as an accidental intruder into the realm of matter; we are beginning to suspect that we ought rather to hail it as the creator and governor of the realm of matter...

But from the first of these 'books' come the grand themes, the meta-narrative of the love of God, forgiveness, salvation, new beginnings and the moral responsibility which should inform and constrain the scientific enterprise.

Not everyone will agree with the position I have argued for here. Individuals must, and do, make up their own minds. But that does not mean that the choice is unimportant. I have already made it clear that the relativist position of '"true" for you but not "true" for me' involves a logical contradiction, not avoided by redefining 'truth'. Truth is the issue at stake. But the matter is not simply an intellectual one, even though this paper concentrates on the cognitive aspects of Judeo-Christian belief. There *is* an affective dimension that concerns the heart, which 'has its reasons of which the reason knows nothing'.[123]

References

Barrow, J. D. (1991). *Theories of everything*. Oxford: Oxford University Press (Vintage).

Brooke, J. H. (1991). *Science and religion: Some historical perspectives*. Cambridge: Cambridge University Press.

Cobern, W. W. (1991). *World view theory and science education research*. NARST Monograph No. 3. Manhattan, KS: National Association for Research in Science Teaching.

Flew, A. G. N. (1970). *Evolutionary ethics*. London: Macmillan.

Hawking, S. W. (1988). *A brief history of time*. London: Bantam Press.

Hooykaas, R. (1957). *Christian faith and the freedom of science*. London: Tyndale Press.

Hooykaas, R. (1966). *The Christian approach in teaching science* (2nd ed.). London: Tyndale Press.

Hooykaas, R. (1972). *Religion and the rise of modern science*. Edinburgh: Scottish Academic Press.

Humphreys, C. J. (1991). 'The star of Bethlehem— a comet in 5 BC — and the date of the birth of Christ', *Quarterly Journal Royal Astronomical Society* (December).

Jeans, J. (1930). *The mysterious universe*. London: Cambridge University Press.

Joad, C. E. M. (1952). *Recovery of belief*. London: Faber and Faber.

Lewis, C. S. (1947). *Miracles*. London: Geoffrey Bles.

National Forum for Values in Education and the Community (1996). *Consultation on Values in education and the Community*. London: School Curriculum and Assessment Authority.

Needham, J. (1969). *The Grand Titration*. London: George Allen & Unwin.

Pascal, B., *Pensées*, trs Cohen, J. M. (1961), Harmondsworth: Penguin.

Phillips, J. B. (1969). *Your God is too small*. London: Epworth Press.

Poole, M. W. (1995). *Beliefs and values in science education*. Buckingham: Open University Press.

Russell, C. A. (1985). *Cross-currents interactions between science and faith*. Leicester: Inter-Varsity Press/ London: Christian Impact.

Ryle, G. (1963). *The Concept of mind*. Harmondsworth: Penguin.

---

[123] Pascal, B. *Pensées*, trs Cohen, J.M. (1961) p. 164, Harmondsworth: Penguin

Soskice, J .M. (1985). *Metaphor and religious language*. Oxford: Clarendon.
Stannard, R. (1982). *Science and the renewal of belief*. London: SCM Press.
Stockmeyer, L. & Chandra, A. (1979). 'Intrinsically difficult problems', *Scientific American, 240*(5) 124
Trigg, R. (1993). *Rationality and science: can science explain everything?* Oxford, UK: Blackwell.

Peter C. Taylor & William W. Cobern

Chapter 10

Towards a Critical Science Education

In the introduction to this book, a critical question was raised: *Whose interests are being served by science education?* Of course, the very asking of this question foreshadows a common dimension to the types of insight one might expect from the ensuing chapters. The question is, indeed, a hallmark of a social constructivist perspective on the generation and legitimation of knowledge, whether by scientists or students of science; a perspective that foregrounds the cultural context within which knowledge is developed, legitimated and made use of.

The viability of a social constructivist perspective is demonstrated throughout this book by the breadth and depth of its explanatory power in addressing and elaborating this question. This is not to suggest, of course, that social constructivism constitutes a homogeneous view of the world. What is clear from a reading of the preceding chapters is that social constructivism, in combination with other theoretical frameworks, can serve a variety of critical social standpoints which illuminate, in complex, intriguing and provocative ways, the various cultural contexts within which both science and students of science are embedded. Our attention is drawn also to the need to understand better the cultural nature of science and its problematic relationship with the broader social values, beliefs and practices of the diverse cultures that are developing science education curricula.

In so doing, the authors propose that a plurality of interests should be served by science education. That many of the authors have evoked the issue of *equity* as a guiding principle for a *critical* science education speaks loudly about the problematic ways in which science education has been practiced historically in many countries, especially the lack of equitable access to a culture-sensitive science education.

In the context of South Africa, Prem Naidoo, Mike Savage and Kopano Taole argue that, in the post-apartheid era, historical inequities in Black Africans' access to quality science education need to be redressed at both societal level (equitable resource distribution) and school level (equitable science curricula). For Black students, science education has an important role in replacing the institutionalised culture of inferiority with a culture of empowerment. New science curricula must be embedded within these students' own (multi)cultural contexts and, in pursuit of the democratic ideal of social justice, must promote questioning and critical reflective thinking.

W. W. Cobern (ed.), Socio-Cultural Perspectives on Science Education, 203–207.

Closely related is Marissa Rollnick's account of the need to redress the inequitable participation of many indigenous students in countries where local languages are valued and practised but where education is conducted, in the interests of economic development, in the tongue of former colonial powers. Because language is believed to mediate both thought and culture, many of these students experience great difficulty in learning science through the medium of a second language, notably English. New science curricula should empower students to participate more fully in science education in accordance with (critical) constructivist views of learning that value the learner's own cultural experiences and that aim to enhance the learner's *voice* and sense of *agency*. Multilingual science classrooms that attend explicitly to the issue of language use and development are proposed as a solution.

For Kate Scantlebury, the patriarchal culture that for too long has favoured the interests of males at the expense of females must be reconstructed in accordance with the democratic principle of social justice. Gender equity in science education can be achieved through *inclusive* curricula that promote the goals, rationalities and everyday cultural contexts of both boys and girls.

The concept of 'culture' that permeates this book is a shared, valued and historically-situated way of living that is characterised by particular socially legitimated ways of feeling, knowing, believing, valuing and relating to others (especially via language). The construct of culture helps to explain major differences between students' home and school lives, between various social groups within a society (e.g., scientists and science teachers), and between groups that differ markedly in language and customs (e.g., Christians and Islamists). But this is not to suggest that culture is static.

Against a seemingly immutable background of traditional cultural practices and symbology, the leading edge of culture comprises a creative and chaotic process of (politically) contested reconstruction. Culture is dynamic. It is at its most vibrant when it is being transformed through the dialogical and dialectical interplay between: self and other; tradition and innovation; past and future; yin (female) and yang (male); freedom and accountability; logos (reason) and eros (feeling); East and West; consciousness and dreaming; materialism and spirituality.

Science, too, is a culture, indeed a curious culture that often masquerades, in the guise of *scientism*, as an *a*cultural activity, one that apparently transcends culture. The image of scientism promotes a distorting ethos of epistemic privilege that, according to various authors in this book, serves to legitimate the 'progressive' social goals of economic development, technological determinism and masculinist privilege, sometimes at the expense of the competing democratic goals of equity and critique; nascent goals that are struggling to find purchase in many countries.

In the face of scientism, local cultures, particularly those in less industrialized countries which import ready-to-use science curricula, are in danger of suffering erosion and loss of integrity as a powerful culture-*in*sensitive science education, operating through the agency of local schools, delegitimates and rapidly displaces traditional ways of knowing, being and valuing. The question of whose interests are being served by this type of science education takes on a particular urgency if cultural diversity and difference is not to give way to an insipid global uniformity based on a distorted image of Western progressivism.

The cultural function of school science has traditionally been to enculturate or assimilate students into the subculture of science. What emerges from this book is, however, a perspective for science education reform called *autonomous acculturation*. Autonomous acculturation is a process of intercultural borrowing or adaptation in which one borrows or adapts attractive content or aspects of another culture and incorporates (assimilates) that content into one's indigenous (everyday) culture. It involves a dynamic view of culture, one that recognises the possibility of and need for inter-cultural adaptation within the rapidly developing global context of social and economic interaction. It involves also a critical view of the cultural nature of modern science, especially the need to deconstruct culture-insensitive myths that perpetuate a distorting and dehumanising image of scientism. And, it involves a dialectical view of the process of cultural adaptation, which recognises the need for reciprocal accommodation of the beliefs, values and practices of modern science and the host culture. The perspective of autonomous acculturation adds the principle of cultural difference and diversity to the democratic ideal of social justice in promoting culture-sensitive science education curricula.

In giving credibility to various critiques of modern science and its role in the development of a global culture, it is important that we guard against an anti-science attitude. Indeed, the authors of this book are united in their valuing of modern science and its potential benefits for personal, social and economic development. Our view is that science education curricula should be designed to acquaint students with science as an important cultural activity for shaping the future of their society, whether it be farming in the local village, conducting medical research in a major institution, or debating the ethics of technological interventions aimed at improving the 'human condition'. We envisage science education continuing its current role of preparing young people as well-informed and discerning personal users of science in their daily lives. Science education is essential for enabling young people to participate in further study in courses that lead to careers as specialist scientists or as public decision-makers concerned with science-related policy issues. But, in the interests of personal autonomy and social responsibility, we cannot afford to pursue this goal alone.

It is equally important for science educators to be critically aware of the tendency toward cultural hegemony of modern science, especially in its scientistic form. One of the self-sustaining (and disempowering) characteristics of culture is its invisibility to its participants. When culture appears to be natural rather than constructed and contingent then inequitable social practices can be experienced as inevitable rather than subject to (political) transformation. Several authors in this book address the way in which scientism restricts science education curricula to serve an unduly narrow range of social and economic interests.

Gili Drori points critically to the global myth of 'science education for development', a powerful international myth which creates an irresistible (and seemingly natural and incontestable) press for local science curricula to serve the pre-eminent social goal of economic development. She argues for a critical and historical analysis of the social discourse embedded implicitly in science education curricula in order to reveal the way that it imprints local cultures (or 'nation-states') with the cultural values of the 'world polity'.

A significant issue for Cath Milne and Peter Taylor is the way that the taken-as-natural culture of scientism is propagated by means of powerful cultural myths which are inscribed in the discursive practices of contemporary science education, and which are sustained from day to day by their uncritical and uncontested (re)telling and (re)practising. This *critical constructivist* perspective on the mythological structuring of experimental science – its naive realist status, the certainty of its knowledge claims, and the value-neutral role of its reporting language – helps us to understand the resilient (and largely invisible) cultural barriers being encountered currently by constructivist science curricula reforms. Additional discursive practices are needed that empower students of science, through critical reflective thinking, to understand the enculturating nature of the science curriculum framework governing (and perhaps unduly restraining) their learning and their agency as learners.

For science education curricula to become *culture-sensitive*, it is important that we understand the extent to which modern science is a cultural form. Several chapters in this book present interesting perspectives on the intimate interconnectedness of culture and modern science and the potential for adapting science to local cultures. Michael Poole's Christian view of modern science illustrates how its historical development was motivated by Biblical views of the orderly and intelligible nature of the physical world (created by a rational, non-capricious God) and led to science's experimental goal of examining nature in order to reveal God's power and wisdom, thereby helping to relieve human suffering. A Christian view of science: rejects anti-religious scientism; recognises that beliefs which underpin the scientific enterprise are congruent with Christian teaching; argues that science is limited in its explanatory power because it cannot serve as a source of moral and ethical principles; and holds that Christian beliefs provide a moral and ethical 'cradle' for nurturing the development of science and providing a context for understanding and valuing science.

Against a background of passionate anti-Westernism critiques by Radical Islamic intellectuals and their attempts to articulate an Islamic Science, Gürol Irzik argues that science education would benefit from an attitude of tolerance and dialogue over the nature of science and its relationship with society. Irzik rejects the positivist image of science, against which Western philosophers of science and religious fundamentalist groups are united in opposition. He proposes a culturally adaptable view of science as comprising: a multiplicity and heterogeneity of scientific disciplines displaying eclectic epistemologies; a notion of scientific objectivity regulated partly by culture-specific values; and a reflexive discourse of self-critique which recognises the necessarily limited role of science in social development.

Drawing on the history of Japanese cultural practices, Masakata Ogawa describes a distinctively Japanese traditional worldview (or cosmology) – Shizen – which involves people in an emotional, spiritual and aesthetic engagement in their natural environment. Ogawa proposes a *multi*-science approach to Japanese school science education, one in which the instrumental practices and products of Western science are interpreted within the Japanese traditional cultural context of Shizen. To observe nature involves much more than a dispassionate recording of events 'out there'; it is to commune with nature. And, to value science is to value more than its ability to explain and predict; its value lies in its ability to enrich the interrelationship between people and nature.

Of course, these tantalising perspectives raise more questions than they do answer, and point to an exciting agenda of research on the culture of science and its relationship with its host cultures. The beliefs, values and practices underpinning modern science are not easily defined once the narrow scientistic view of science has been rejected (perhaps scientism's readily definable nature helps to explain its popularity?). Although the extent of the interconnection between the culture of modern science and its host cultures remains unclear, a number of authors argue that specific cultural (moral, spiritual, aesthetic) values should be incorporated into the self-regulating ideal of objectivity, a standard which, more than any other, is associated with the nature of modern science.

So, in the light of cultural pluralism, and bearing in mind the increasing credibility of relativism within a postmodern worldview, we need to address the question of what standards should govern the cultural adaptation of the eclectic practices of modern science? Also, bearing in mind the difficulty of defining the beliefs and values of any social group, especially in multicultural societies where competing sets of beliefs and values vie for an increasing share of finite resource allocation, how do we answer the question of the extent to which these standards should be culture-sensitive?

While these issues are being debated well into the next century by scientists, philosophers of science, and science educators, we wish to re-emphasise the emergent theme of this book: if science education is to 'come alive' for students of all cultures then it is important that science be experienced as a process of critical enculturation. We join Kopano Taole in calling for a critical science curricula that empowers students to:

- develop a sensitivity to and an appreciation of the natural sciences as a value laden human enterprise;
- recognise and acknowledge contributions to the natural sciences by different cultures, religions and societies; and
- identify and deal with biases and inequities implicit in and imported through the natural sciences.

Thus, a critical science education offers the empowering prospect that students will learn to adapt their local cultures to scientific ways of knowing, believing and valuing, AND learn to adapt science to their own cultural ways of knowing, believing, and valuing.

About the Contributors

William W. Cobern is an associate professor of science education at Western Michigan University, Kalamazoo, Michigan. He received his undergraduate degree in biology and chemistry from the University of California, San Diego (Revelle College) in 1971. After teaching high school science for several years he studied for his doctorate in science education at the University of Colorado, Boulder receiving his Ph.D. in 1979. Dr. Cobern then spent five years as a faculty member at the University of Sokoto, Nigeria. From this basis in experience he developed a worldview theoretical framework for conceiving of science as an aspect of culture which warrants the assertion that meaningful science learning only occurs to the extent that scientific knowledge can find a cognitive niche within the everyday thinking of ordinary people. Dr. Cobern has published articles in *Science Education*, the *Journal of Research in Science Teaching, Science & Education*, and *Perspectives on Science and Christian Faith*. He is also section editor for Culture and Comparative Studies in the journal *Science Education*.

Gilis Drori, a native of Israel, completed her master's degree studies at Tel Aviv University and published several papers in Israeli academic journals. She recently completed her doctoral degree (1997) at Stanford University's Department of Sociology. Her interest is the globalization of science and technology and the patterns of diffusion of policies and programs of science and science education. Currently her work focuses on the ways by which such globalization and diffusion processes alter the nature of nation-states and their practices, through the introduction of science-related, and other, Western models of understanding. She is most interested in Israeli science policy issues, and is currently researching the innovative Israeli strategy for technology and industry linkage through what are known as "technological incubators".

Gürol Irzik is an associate professor of philosophy and chair of the Philosophy Department at Bogazici University, Turkey. He came to philosophy with a background in electrical engineering (BS/1977) and mathematics (MA/1979). His doctoral degree is in the history and philosophy of science (Indiana University-Bloomington, 1986) and in 1995-96 he was a visiting fellow at the Center for Philosophy of Science at the University of Pittsburgh. Dr. Irzik's areas of specialization are questions of methodology and the relationship between positivist and post-positivist approaches to science; philosophy of social sciences with special emphasis on making causal inferences on the basis of statistical data; and theories of causation. He has published in such journals as *Philosophy of Science, British Journal for the Philosophy of Science, Studies in History and Philosophy of Science*, and the *Philosophy of Science Association Proceedings*.

Cathrine E. Milne is an Australian with 18 years of teaching experience at both secondary and tertiary levels. Her secondary science teaching experience was in the Northern Territory, Australia. Currently, she is studying for her Ph.D. at Curtin University of Technology in Perth, Western Australia. Ms Milne's research interests lie in the examination of school science as a culture and the implications that has for course development and classroom interactions. In particular, she is interested the interaction between science cultural myths and the practices and discourses of school science.

Prem Naidoo is the Director of the African Forum for Children's Literacy in Science and Technology based at the University of Durban-Westville, South Africa. Previously, he was a lecturer in science education and the Director of the Education Policy Unit at the University of Durban-Westville. He was one of the principle organizers and promoters of the first Africa Science Teacher Education meeting held 1995. This meeting sought to bring together science teacher educators from all of Africa. Mr. Naidoo has a long-standing interest in promoting science education among disadvantaged peoples.

Masakata Ogawa, an associate professor of science education at Ibaraki University, is a graduate of Kyoto University where he received the doctoral degree in plant physiology (1982). "Science as a foreign culture" is the key idea in his research. He has conducted cultural studies in science and technology education and STS education. He has published many articles, chapters and books. His work has appeared in *Science Education* and the *International Journal of Science Education*. Recently he organized an international joint research project, "Effects of Traditional Cosmology on Science Education", funded by the Ministry of Education, Sports and Culture, Japan.

Michael W. Poole taught physics at a London comprehensive school before undertaking work on the preparation and broadcasting of a series of radio programs on science-and-religion for overseas audiences. He became lecturer in science education at King's College London in 1973 with research interests in the interplay between science and religion, with special reference to its educational context. He is currently Visiting Research Fellow in the School of Education at King's College. In 1993 he was made a Fellow of the Royal Society of Arts and in 1995 received a Templeton Award for his MA module in science and religion. He received a further award in 1996 for his paper, 'A critique of aspects of the philosophy and theology of Richard Dawkins', *Science and Christian Belief*, **6** (1) 41-59(1994). His books include *A Guide to Science and Belief*, Oxford: Lion Publishing (1994) and *Beliefs and values in science education*, Buckingham: Open University Press (1995). He is also the author of some forty papers and articles on issues of science and religion.

Marissa Rollnick is presently a senior lecturer in the chemistry department at the University of Witwatersrand. Her primary responsibility is coordinating and teaching courses to under prepared student most of whom are second language speakers of English. Prior to 1990 she worked in Swaziland for fifteen years where she worked both in a teachers college and at the University. It was here that she first carried out research on language in science education and this was the subject of her Ph.D. work. Since returning to South Africa, her home, she has participated in National Curriculum committees and has chaired the Southern Africa Association for

Research in Mathematics and Science Education. She has also done further research into other aspects of language issues in science education.

Michael Savage is the Technical Advisor for the African Forum for Children's Literacy in Science and Technology and is based at Chancellor College, Malawi. He is one of Africa's most distinguished science educators and along with Prem Naidoo helped to bring about the first meeting of the Africa Science Teacher Education meeting held 1995 at the University of Durban-Westville, South Africa.

Kathryn Scantlebury is the Secondary Science Education Coordinator and Associate Professor in Chemistry and Biochemistry, with a secondary appointment in Women's Studies and a cooperating appointment in Educational Development, at the University of Delaware. Dr. Scantlebury has extensive teaching and research experience in gender issues and science teacher education. She has taught undergradaute courses in methods and chemistry, graduate courses in science education and gender studies, supervised student teachers. Dr. Scantlebury has also taught chemistry to elementary preservice and inservice teachers, run workshops for teachers on gender equity in science and served as a consultant for numerous schools and districts on issues related to gender and education. In addition she has worked extensively with non-profit groups such as the Girl Scouts, Girls Inc. and AAUW. She co-edited the monograph, "Science 'Coeducation': Viewpoints from Gender, Race and Ethnic Perspectives" and has published in the *Journal of Research in Science Teaching*, *Journal of Science Teacher Education*, *Journal of Chemical Education* and *Journal of Women and Minorities in Science and Engineering*.

Kopano Taole is the program manger for Science, Engineering and Technology (SET) Education and Awareness Programmes at the Foundation for Research and Development, Pretoria, South Africa. His current job responsibilities are directed toward equity issues in the development of SET research expertise in South Africa. Dr. Taole holds a doctorate of education in mathematics education from Columbia University, USA, and has over 20 years of experience in mathematics and mathematics education. He is the author of several mathematics textbooks for both primary and high school. His recent research work in gender in education was published in *Educational Studies in Mathematics*.

Peter C. Taylor is presently a senior lecturer and science teacher educator at the National Key Centre for School Science and Mathematics, Curtin University of Technology in Perth, Australia. His teaching and research focus mainly on ways of empowering teachers to transform the culture of their science and mathematics classrooms. He draws on anthropology, constructivism, feminism, postmodernism, literary theory, history and philosophy of science, and critical theory as contributory referents for making sense of teachers' professional activities, and for constructing alternative visions of a more equitable and inclusive society and education system. His recent publications have explored the historio-cultural myths that govern the curriculum practices of modern education. Dr. Taylor is a member of the editorial boards of the *Journal for Research in Science Teaching* and the *Journal of Science* *and Instructional Technology*. He is a regular participant in annual meetings of the American Educational Research Association and the National Association for Research in Science Teaching (NARST). In 1997 he received the NARST Distinguished Early Career Achievement Award.

Index

Science & Technology Education Library

Series editor: Ken Tobin, *Florida State University, Tallahassee, Florida, USA*

Publications

1. W.-M. Roth: *Authentic School Science.* Knowing and Learning in Open-Inquiry Science Laboratories. 1995 ISBN 0-7923-3088-9; Pb: 0-7923-3307-1
2. L.H. Parker, L.J. Rennie and B.J. Fraser (eds.): *Gender, Science and Mathematics.* Shortening the Shadow. 1996 ISBN 0-7923-3535-X; Pb: 0-7923-3582-1
3. W.-M. Roth: *Designing Communities.* 1997 ISBN 0-7923-4703-X; Pb: 0-7923-4704-8
4. W.W. Cobern (ed.): *Socio-Cultural Perspectives on Science Education.* An International Dialogue. 1998 ISBN 0-7923-4987-3; Pb: 0-7923-4988-1

KLUWER ACADEMIC PUBLISHERS – DORDRECHT / BOSTON / LONDON

Science & Technology Education Library

S[illegible] Editor: Ken Tobin, Florida State University, Tallahassee, Florida, USA

Publications

1. W.-M. Roth: Authentic School Science. Knowing and Learning in Open-Inquiry Science Laboratories. 199[illegible] ISBN [illegible]-3088-9; Pb 0-7923-3307-1
2. L.H. Parker, L.J. Rennie and [illegible] (eds.): Gender, Science and Mathematics. Shortening the Shadow. [illegible] ISBN 0-7923-3535-X; Pb 0-7923-3582-1
3. W.-M. Roth: Designing Communities. 199[illegible] ISBN 0-7923-4703-X; Pb 0-7923-4704-8
4. W.W. Cobern (ed.): Socio-Cultural Perspectives on Science Education. An International Dialogue. 1998 ISBN 0-7923-4987-5; Pb 0-7923-4988-3

[illegible] ACADEMIC PUBLISHERS – DORDRECHT / BOSTON / LONDON

9 780792 349884